Thinking
A Guide to Systems Engineering Problem-Solving

Thinking
A Guide to Systems Engineering
Problem-Solving

Howard Eisner

CRC Press
Taylor & Francis Group
Boca Raton London New York

CRC Press is an imprint of the
Taylor & Francis Group, an **informa** business

CRC Press
Taylor & Francis Group
6000 Broken Sound Parkway NW, Suite 300
Boca Raton, FL 33487-2742

International Standard Book Number-13: 978-0-367-11219-6 (Paperback)
International Standard Book Number-13: 978-0-367-11220-2 (Hardback)

Library of Congress Cataloging-in-Publication Data

Names: Eisner, Howard, 1935- author.
Title: Thinking : a guide to systems engineering problem-solving / authored by Howard Eisner.
Description: Boca Raton : CRC Press, Taylor & Francis Group, 2019. | Includes bibliographical references.
Identifiers: LCCN 2018043908 | ISBN 9780367112196 (pbk. ; alk. paper) | ISBN 9780367112202 (hardback ; alk. paper) | ISBN 9780429025365 (e-book)
Subjects: LCSH: Systems engineering. | Problem solving.
Classification: LCC TA168 .E3875 2019 | DDC 620/.0042019—dc23
LC record available at https://lccn.loc.gov/2018043908

Visit the Taylor & Francis Web site at
http://www.taylorandfrancis.com

and the CRC Press Web site at
http://www.crcpress.com

This book is dedicated to my wife, June Linowitz, who has spent many years of her life as an artist. I have had the pleasure of seeing how she does "artist projects" and have found, invariably, that it takes a lot of thinking as well as problem-solving. So she knows a lot about the processes and perspectives of this book, just by "doing," and "experimenting". Nothing like trying and persevering.

Since June is a sculptor, she has had to teach herself how to "visualize." I found, as I was researching for this book, that Einstein declared this skill as critical in his scientific explorations. Interesting as to how certain skills are needed, even as they come together from different fields.

I also dedicate this book to two of my children, Oren and Susan, and their spouses Tara and Joseph, and their children Zach, Ben, Becca, and Gabe. And I don't want to miss yet another grandchild, Jake. It is my hope that one day they will pick up this book, read it cover to cover, having fun all the way.

Howard Eisner
Bethesda, Maryland

Contents

Preface

From my own experience, as well as the literature, it appears that there are many different ways to approach problem-solving, and therefore, many ways of thinking. This book explores a variety of these ways, to include

Systems thinking	Inductive thinking
Design thinking	Deductive thinking
Lateral thinking	Out-of-the-box thinking
Disruptive thinking	Fast and slow thinking
Critical thinking	Reductionist thinking

It also briefly looks at some of our great thinkers, with the notion that we can find out more about thinking by what they did and, in some cases, their writings.

It also briefly examines thinking in a group or team situation, which can be highly functional as well as dysfunctional (sometimes called groupthink). Improving the thinking in group situations is left as a "problem to be solved;" perhaps, the new thinking patterns will help in this regard. I suppose one should have such expectations.

A basic reason for this overall exploration is to (hopefully) assist the engineer by enhancing his or her approach to problem-solving in their area of expertise—that of designing and building large-scale complex systems. Yet another hope is that it will assist all serious problem-solvers, whatever their domain of inquiry.

Both the domains of thinking and that of systems engineering have a rich literature. However, it seems that information on the connection between these two worlds is relatively sparse—hence, this treatise. So, if the engineers of today can open their minds (and hearts) to some new ways of thinking, better approaches and solutions might be found.

As we approach this broad and most interesting topic, we find that there is an international trend that must be cited and acknowledged. That is, the field known today as STEM—science, technology, engineering, and mathematics—is the 800-pound gorilla in the room. That is, across the world, there is more and more emphasis on the need for STEM as well as how we might (or might not) meet that need. Businesses and government generate much of that need, and schools figure out how to satisfy that need. And one subject that appears to be behind all of STEM, all of its individual pieces, is simply "thinking." We require better and better thinking to get us where we need to be in STEM. So, one might say that the primary ingredient in STEM as we move along both nationally and internationally is an improvement in how we approach and think about problems and how they are to be solved. This is a nontrivial task. So let's start now.

Author

Howard Eisner, PhD, spent 30 years in industry and 24 years in academia. In the former, he was a working engineer, manager, executive (at ORI, Inc. and the Atlantic Research Corporation) and president of two high-tech companies (Intercon Systems and the Atlantic Research Services Company). In academia, he was a professor of engineering management and a distinguished research professor in the Engineering School at the George Washington University (GWU). At GWU, he taught courses in systems engineering, technical enterprises, project management, modulation and noise, and information theory.

He has written seven books that relate to engineering, systems, and management. He has also given many lectures, tutorials, and presentations to professional societies (such as INCOSE—International Council on Systems Engineering) and the Osher Lifelong Learning Institute (OLLI). In 1994, he was given the outstanding achievement award from the GWU Engineering Alumni.

Dr. Eisner is a life fellow of the IEEE and a fellow of INCOSE and the New York Academy of Sciences. He is a member of Tau Beta Pi, Eta Kappa Nu, Sigma Xi, and Omega Rho, various honor/research societies. He received a (Bachelor's of Electrical Engineering (BEE) from the City College of New York (1957), an MS degree in engineering from Columbia University (1958), and a doctor of science from GWU (1966).

Since 2013, he has served as professor emeritus of engineering management and a distinguished research professor at GWU.

chapter one

Aspects of thinking

We observe the world around us and try to figure out how it works. We do this not only for survival, but also to improve our quality of life. To succeed in these endeavors, we have had to think. And, so far, we've been able to do this. Some of it comes "naturally"; for some of it, we've had to work hard to master its various elements. And as sophisticated as we are, at its root, it's still about survival and quality of life.

It all seems to start with *observation*. That is, we notice something and then ask—what's going on here? In the early days, an easy target for observation was the skies. We saw parts of our solar system and the numerous stars and developed *"hypotheses"* about the underlying phenomena. Some of these were right, while some were wrong. But we continued to observe, formulate hypotheses, and *test these hypotheses*. The latter were formalized into sets of equations and "models." Observe and test, observe and test, until the two lined up. And when they did, we had a "law" that we could depend upon. Finding the "law" that worked with the observable data was the process as well as the goal. And we've been doing that for years and years. We've been *thinking* about all of it for years and years. So the steps have been

- a. Make observations
- b. Develop hypotheses, and
- c. Test these hypotheses.

The "thinking" part appears dominant in the second step. That's the part that asks and tries to answer the question—what's going on here?

Definition—thinking

We can find a starting point in the dictionary, as follows [1]:

Thinking definitions

1. To form or have in the mind
2. To hold in one's opinion
3. To believe
4. To determine

5. To purpose
6. To bring to the mind

Other words that are used in and around the formal definition are to

1. Reason
2. Cogitate
3. Reflect
4. Speculate, and
5. Deliberate

We will elaborate upon this and other notions regarding thinking, throughout this introduction and book.

Reasoning

Also, we have some intuitive perspective about "reasoning," figuring that it comes into the process in some way or other. Reasoning involves taking a step-by-step approach to thinking. One step leads to and infers the next, and so on until we have a sequence of ideas that seem to build upon one another. It is also connected to *"logic"* in that we are precise and rigorous in our approach. This is, as many know, in consonance with Spock's approach to the universe. For the sake of completeness and further elucidation, we look here at the dictionary definition of "reasoning":

Reasoning definition [1]
1. To think coherently and logically
2. To argue, conclude or infer
3. To support and justify
4. To persuade

We also have some "tools" that help us think. These are induction and deduction, both of which are briefly examined here.

Induction

Another definition here is simply [1] "reasoning from particular facts or individual cases to a general conclusion." There is also a "mathematical" approach to induction that we learned in high school. This may be expressed as

- Prove that if a proposition is true for "n," then it is also true for "(n + 1)"
- Show that the proposition is true for (n = 1)

Possibly, this can be better understood by visualizing a line of soldiers. The first of the earlier one says that we must show that, if a soldier falls, then the soldier to his immediate right must also fall. Then, one shows that the first soldier falls. One can see then how all the soldiers must fall. This form of reasoning and observation argues from the specific to the general, and is absolutely rigorous.

Deduction

Here again, we can look at the dictionary to understand at least a couple of definitions of this process:

- Reasoning from a known principle to an unknown
- From the general to the specific

So, we start out with a generally accepted proposition and then declare that a specific statement is true. Three examples are [2]

1. To earn a master's degree, a student must have 33 credits. A student named Stuart has 35 credits so he will earn a master's degree
2. All birds have feathers, and finches are birds, so they have feathers
3. All cats have an acute sense of balance. Kitty is a cat and therefore she has an acute sense of balance.

Educating to think

We often claim that our education system is built upon the idea that we are teaching the young ones to think and not to memorize a set of facts, however interesting these facts might be. It is "easier" to teach facts, but is much less productive. It is difficult to teach people how to think. One must use concrete examples of how it is done, and generally, must accept a wider range of "solutions." This remains one of the problems of teaching that will stay in front of us as an issue to deal with and to ultimately solve. Hopefully, we can make progress, year by year, in this somewhat elusive problem.

Problem-solving

We certainly have to think when we are trying to solve a problem. However, we may be thinking (about a wide variety of things) and not engaged in problem-solving. Here are some of the steps we may have to address in solving a day-to-day problem [3]:

a. Define the problem in precise terms
b. Identify the key factors that relate to the problem

    c. Develop a set of inferences from the problem itself and for each of
       the key factors
    d. Define a group of alternative solutions (more than one)
    e. Select the "best" solution

There is clearly a lot of thinking for each of these steps to be completed
satisfactorily.

Systems engineering

This book is also intended to connect directly to the world of "systems
engineering." The latter is a formalized set of procedures and steps that
we have devised to successfully build complex systems. How does it
do this? It recognizes that systems engineering has about 30 elements,
and various steps are required so as to execute each of these elements.
And they are not automatic; each step requires "thinking," and lots of
it. So in that sense, "thinking" is a foundation or base for carrying out
systems engineering. And systems engineering is a foundation or base
for large-scale (systems) problem-solving. The pieces fit together: systems
engineering, problem-solving, and thinking.

In view of the "position" and importance of systems engineering,
several chapters deal with systems engineering in an explicit way. To be
more precise:

- Chapter three deals with "suggestions for the systems engineer"
- Chapter seven cites "special thoughts" from selected engineers, and
- Chapter eight elaborates upon a "top dozen" for systems engineers

The better idea

Another domain in which we see special thinking has to do with finding
the "better idea." And this occurs in all fields, with numerous intrinsic and
extrinsic motivations. Let us briefly cite four such cases dealing with [4]

- Copernicus
- Gandhi
- Darwin, and
- Einstein

Copernicus (1473–1543)

Around the turn of the 16th century, the prevailing notion was for an
earth-centered system in the world of astronomy. It was Copernicus that
argued ultimately for a heliocentric or sun-centered approach. Apparently,

he deepened his levels of observation and found that these data did not support an earth-centered view. Better observations—better inferences. But it was Copernicus that made it all happen, producing documentation in the form of a classic treatise [5]. Now that was a real breakthrough above and beyond just standard superior thinking.

Gandhi (1869–1948)

Gandhi was an original thinker, dealing with the oppressive relationship between his country, India, and Great Britain. He was committed to the principles of "service," which he built upon to develop a philosophy of what many now call "passive resistance." He was able to go beyond the principle and into the practice, an extremely difficult path to take. But he had the deep conviction as well as the discipline to make it all happen in the face of great odds. His pioneering approach laid the foundation, in effect, for what Martin L. King was able to achieve in the United States.

Darwin (1809–1882)

Darwin claimed to be a modest man but with a rather powerful devotion to science and a keenness of observation about the world around him. He was able to formulate what was called a new "theory of evolution by natural selection." The short form of this phrase also became "survival of the fittest." During a very productive century of original thinking, Darwin stood out as a man who put the pieces together in a most extraordinary manner. Here again, we see the sheer power of curiosity as well as that of observation.

Einstein (1879–1955)

With Einstein, we reach what many have called the supreme icon in the world of science, physics in particular. He formulated the theories of relativity that have stood the test of time and exhaustive experimentation. Curiously, he was not first in his class and held a modest position in a Swiss patent department in Bern. He left Europe, came to the States, and continued his research and studies at Princeton's Institute of Advanced Studies.

What did they have in common?

For this author, the four "mortals" cited earlier had four attributes in common:

1. A love of their chosen fields
2. Absolute persistence

3. Superior intelligence
4. An ability to challenge conventional wisdom

Above all, the last item given earlier allowed them to make their extraordinary observations and contributions.

The results of their works are very clear now, but in many cases, it is still not completely clear as to what their thinking was, in detail. As applied to some superior persons, we know what they thought since they documented their approach. But for many, we see the results but not the thinking process. So be it. It gives us more of an incentive to investigate not what they did, but what they were thinking. And that's the more difficult part of this treatise.

Business-oriented better ideas

A significant amount of time and energy is devoted to finding better ideas in the "business" world. These may not measure up to the worlds of science and medicine, but hundreds of millions of dollars may be at stake every year. Here is a short sampling of better ideas that have come out of the domain of business.

Peter Drucker

Peter Drucker, a professor, was the ultimate guru in the business world. He documented his ideas in many books, and gave seminars and lectures for the business community. He was a leader in the field of business education, devoted to improving conditions and activities in that domain. An example of just one of his contributions was that of "Management by Objectives" (MBO). This idea, very much accepted in the business community, focused on identifying objectives for management, and supported that approach with specific measurement schema. This author worked under such a "system" and found that it worked in the real world.

Federal Express

The success of Federal Express, and other mail delivery service companies, was very much dependent upon having the most efficient and cost-effective methods for handling their packages. The "literature" had many models of the so-called transportation problem, but the bottom line was that Federal Express brought their mail into a central hub (Memphis) overnight and then fanned the mail to destinations from there. This was, and remains, a better idea that went from "thinking" to superior implementation, day after day.

Netflix

So there came to be a time when Reed Hastings went to see Blockbuster, suggesting a better way to deal with Video Home System (VHS) video tapes. Apparently, Blockbuster rejected his approach, so he went off to try the new idea on his own. He implemented a scheme for mailing DVDs and streaming videos that literally forced Blockbuster into bankruptcy. This remains a rather spectacular story of how the better idea can take hold and lead to the creation of a quite significant company.

Other concrete steps involved in the thinking process

As we complete this introductory chapter, we look briefly at some specific and concrete steps one can take to help in the thinking process. If one is trying to facilitate such a process, one might start with a series of questions. The first such question, in the context of problem-solving or finding the "better idea," is:

What is it that we know (about the problem)?

This is an explicit search for the known body of knowledge and its clear articulation. It is a statement, if you will, regarding the existing state of the art (or of the science, etc.). It's a statement about where we are today. Of course, we should be able to answer such a question in concrete terms.

From there, we move on to the next logical question:

What is it that we don't know?

This is an attempt to be clear and complete about the specific pieces of information that we do not have but need to know. Examples might be

1. We need to know more about how the economy will react to a stimulus of size X.
2. We need to know how people will vote in the next election, given various scenarios.
3. We need to know how this substance will affect the heart's behavior when applied every day with dosage Y.
4. We need to know how to eradicate these specific types of cancer cells.

Clearly, the implication is that these are the key questions to make progress with the problem at hand. And then we can move beyond that with:

*What will happen when we get the answers to
the "what we don't know" questions?*

There are several "answers" to that seemingly innocuous question. Some of them are

1. We will have solved the problem.
2. We will have taken an important step in the direction of a solution.
3. We very likely will have deepened our understanding of the problem.
4. We will be closer to developing the true costs of moving forward.

Remember, we are not talking about trivial problems. Mostly, we are looking at significant problems and issues that take years and years of study by many (numerous) competent people. Think about the National Cancer Institute and the work that they do, each and every year, with quite significant funding. So, there's a lot at stake, and a lot therefore to pay attention to in terms of thinking patterns and how to be successful with the process.

Flaws in thinking that need to be addressed

As we complete this introductory chapter, we wish to comment on the "other side of the ledger;" namely, the fact that there are several hurdles to deal with in terms of thinking and setting forth "the better idea." Some of these hurdles are in the domain of group processes that are set in motion to solve a problem. Once we move into groups, we have to reckon with the dynamics of a group, which at times can be quite dysfunctional. An overview word for such behavior has been called groupthink [3]. This is a largely derogatory notion, suggesting that there are times when groups puzzle their ways through problem areas and mostly fail miserably. These are obstacles to achieving superior thinking, but they ultimately need to be reckoned with. Further discussion of this area can be seen in later chapters of this book.

References

1. *"Webster's New World Dictionary"*, Second College Edition, Prentice Hall Press, 1986.
2. examples.yourdictionary.com.
3. Eisner, H., *"Managing Complex Systems: Thinking Outside the Box"*, John Wiley, 2005.
4. Gelb, M. J., *"Discover Your Genius"*, HarperCollins, 2002.
5. Copernicus, N., *"The Revolution of the Heavenly Spheres"*; Great Books of the Western World, Volume 16, Encyclopedia Brittanica, Chicago, 1952.

Exercises

1.1 What are five words, other than those in the book, that bring to mind the meaning of "to think"?

1.2 What are the differences between "to reason" and "to think"? Are they just about the same?

1.3 Identify five scientists, and their theories, other than the four cited in this chapter.

1.4 Identify five business-oriented "thinkers (persons)", other than those in this chapter, and their key thoughts/ideas.

1.5 Identify five companies, other than those cited in this chapter, that are associated with key thoughts/ideas. What are/were these thoughts/ideas?

chapter two

Thinking outside the box

Introduction

This chapter looks at what has been called "thinking outside the box," a term that has come into common use in the last 15 years or so. Its meaning? Just what one might expect. It is taking a point of view that is relatively rare and outside today's norms. It is considering a problem from a different angle. It is bypassing the thoughts that might be held by most people and coming up with ideas that are unusual. As interpreted in this treatise, it is what perhaps 5% of the people might think, in distinction to the view held by 95% of the vast majority.

This author, some years ago, wrote a book [1] whose subtitle was "thinking outside the box." This chapter reiterates some of that material. However, much of this text is new and considers viewpoints advanced by the author in the past several years.

A short note that tells a story

We start out with an example of thinking outside the box that makes it clear as to what might be outside the box and also how important, in terms of potential impact, such thinking might be.

In the early part of the 20th century, the key railroad folks in the country got together to do some strategic planning. They finally addressed the question—what business are we in? —which is a typical matter to be considered in any strategic planning discussion. Their answer was a resounding:

We are in the railroading business, of course

Now that answer, accepted by all, tells the story as to why none of the railroad people were key drivers in the upcoming "air transportation" business. They could have said they were in the "transportation business," in which case many of those in the room could have targeted airplanes and aviation. But they didn't. So by failing to broaden their view of the business they were in, they shut their thinking to the new opportunities in the new technology and business of flying. Of course, this could have had major impacts in terms of the successes and failures of several companies.

What's inside and what's outside?

One way to look at thinking outside the box is to contrast and compare what's inside with what's outside. Let's look at the following listing.

What's inside the box	What's outside the box
1. Using the "man-month" for planning purposes is entirely appropriate	Different skill levels need to be considered for each man-month
2. Very high performers in your company are not concerned about administrative matters	Very high performers tend to be concerned about everything
3. As a general principle of management, measure everything you can think of	Measure only those parameters and variables that "tell the story"
4. The customer is always right	There are many times when the customer is wrong
5. Overpromise so as to improve ratio of wins to losses	Try to underpromise and overdeliver
6. Consistently have your people work overtime to improve profitability	Go to overtime when needed to catch up
7. Avoid serious and detailed planning to maintain flexibility of action	You need a plan to set direction as well as communicate with all personnel
8. Accept all requirements from the customer as fixed and inviolate	There are times when certain requirements need to be challenged and negotiated
9. There is usually no need for a formal risk assessment for your projects	Selected projects need to undergo a formal and detailed risk assessment
10. Rely more on your staff people than your line people and organization	Your line people have priority for communication and listening
11. Integrate all stovepipes	Integrate only when it's cost effective to do so
12. Don't improve your product if it is selling as it is	Always look at product improvement possibilities

These are just a dozen areas in which inside and outside the box may be contrasted. In some cases, the comparison may be challenged. But by and large, the "outside" considerations are the less than usual approaches. Take a moment, as the reader, to think about what happens in your company. Can you think of other areas in which comparisons may be usefully constructed?

Eight suggestions

We come now to citing and briefly discussing quite specific ways to "think outside the box." These are general, and apply broadly to several fields. These eight suggestions are based upon those defined by this author in a previous work [1].

Broaden and generalize

The earlier "story" regarding strategic planning by the railroad executives is a good example of broadening, or not broadening, the answer to the question— what business are we in? By broadening, we are taking a more inclusive approach. In that way, we are potentially including new and better solutions to the problem(s) we are facing.

Crossover

This approach has to do with the following: if you have produced (built) a system for one customer, try to provide this system to other (crossover) customers. An example might be that if you've just built a personnel tracking system for the Army, try to provide this same system to the Navy and the Air Force. Today, for various reasons, this approach is not taken. But if it is, it is likely that huge benefits in cost and time will be achieved.

We can also extend this approach to state and local governments, and crossover from the government customers to commercial customers. These approaches are creating new forms of "leverage," which is highly desirable. So the generality here is that we should be looking everywhere for new opportunities to create new leverage, built upon our current business base.

Question conventional wisdom

Much of the progress made by significant researchers has been achieved by questioning the conventional wisdom at that time and place. This goes back a long way, as in the cases of

a. Challenging that the earth is flat
b. Moving from an earth-centered system to a sun-centered system "model"

Moving to more modern times, we would cite as examples the following technical and management people who were able to successfully challenge the conventional wisdom of their time [1]:

Hewlett and Packard	Igor Sikorsky	Robert Oppenheimer
Andy Grove	Robert Goddard	Ross Perot
Norman Augustine	Admiral Rickover	Bill Gates
Lee Iacocca	Jonas Salk	Steve Jobs

We do understand that these and other pioneers in various fields, at one time or another, have taken to challenging the prevailing wisdom, and thereby making the progress than we have come to see and appreciate.

Back of the envelope

This suggestion has to do with finding tentative solutions to problems by trusting yourself (and others) to use their intuitions, to the maximum. The claim here is that senior people have years and years of problem-solving experiences. Based upon these experiences, they have also developed sharp and accurate intuitive approaches. These approaches can yield preliminary solutions, to be confirmed by more detailed analyses down the road.

This notion is supported by the fact that many solutions to quite complex problems can literally be expressed on the back of an envelope. Perhaps, the classic example is the familiar $E = mc^2$. Here are some others:

a. Newton's gravitational law (& formula)	d. Ohm's law
b. Shannon's definition of information	e. Boyle's law
c. Maxwell's electromagnetic equations	f. Charles' law

Expanding the dimensions

A direct example of expanding the dimensions is the search for a grand unified theory of physics. As it turns out, a "solution" was found by expanding to ten dimensions [1]. Although ten dimensions are not physically realizable, we see the power of being able to investigate what might happen in ten-dimensional space.

Another example is taking the two-dimensional spreadsheet that we had come to know and love (Excel, Lotus 1–2–3, etc.) and expand it to three dimensions. I first had this experience on a trip to Boeing. A colleague there said, "How would you like to see BoeingCalc? It's our 3D version of a spreadsheet." This was not a massive breakthrough in technology, but it served a specific purpose and clearly expanded the dimensionality of the spreadsheet.

Remove constraints

When thinking about solutions to the problems we are dealing with, there is always the question of rejecting answers due to constraints that we perceive to be in effect. Some of these constraints are simply

- Not enough money to allow us to do x or y or z
- Not enough time for us to do x or y or z
- Inadequate facilities to do "vacuum thermal" testing
- Insufficient senior personnel to apply to the set of tasks
- Insufficient top-level management commitment and allocation of resources

For each of the earlier constraints, we might well consider how the constraint might be removed, rather than accepting the constraint as a "given." As examples,

a. Have we actually requested more money and time?
b. Have we looked in detail at our schedule and tried to do more tasks in parallel?
c. Have we explored the use of subcontracting out our vacuum thermal testing?
d. Have we tried to ramp up our hiring of senior people, or moved some of our senior personnel from other projects to the current project?
e. Have we asked top management to allocate resources in a different way, including new data to support that argument

Out-of-the-box thinkers tend to find a way to remove apparent constraints, leading to new and sometimes astonishing successes.

Thinking with pictures

This approach can be made clearer by the following:

- Consider describing in words what the Mona Lisa looks like to a friend vs. giving him or her a picture of Mona Lisa
- Consider describing in words what an optical illusion looks like vs. showing a picture or sketch of that optical illusion
- Consider describing in words what the Department of Defense (DoD) acquisition process looks like vs. providing a diagram of that process

We can easily see that a picture, or a sketch, or a diagram, will provide a lot more information in distinction to a linear set of words or text.

It was Rudolf Arnheim who suggested that "visual perception is a distinctly cognitive activity" [2]. By the very title of his book (Visual Thinking), we can infer that the professor saw visuals as an important aspect of thinking. And that is indeed an important point that he made. Others have also reinforced that idea. So has the oft-used expression "one picture is worth a thousand words."

Systems thinking

The final element of "out-of-the-box" thinking is that of systems thinking. This will be described in considerable detail in chapter three. Here we provide an overview, with some of the main points of this approach. It tends to involve

- Taking a broader and more holistic approach to problem-solving and thinking
- Integrating the basic elements of thinking
- Seeing the relationship between intermediate steps and the whole
- Emphasizing features of the whole vs. the pieces and parts of the whole
- Seeing the "larger picture"

Now we move on to the next chapter dealing with many more aspects of systems and thinking.

References

1. Eisner, H., *"Managing Complex Systems: Thinking Outside the Box"*, John Wiley, 2005.
2. Arnheim, R., *"Visual Thinking"*, University of California Press, 1969.

Exercises

2.1 What, for you, are five thoughts "outside the box"? Why? Explain in detail.
2.2 Twelve persons are cited in this chapter. How did they question conventional wisdom?
2.3 Cite three ways that you suggest "expanding the dimensions" of a problem. Be specific.
2.4 Other than those mentioned in this chapter, name five well-known "thinkers," what they thought about, and how and why they were successful in their thinking.

chapter three

Systems and thinking

This chapter is about "systems"—what they are, thinking about them, and the relationship between thinking and systems.

What is a system?

A short but useful definition of a system is simply [1]: it is an assemblage of elements or components that are interconnected so as to harmoniously perform one or more functions, to serve one or more purposes. We expand upon that definition and recognize that a system can also be a procedure for accomplishing a task. And finally, a system can be one or more bodies of knowledge that explain how various parts of nature or man-made systems work. Generally, there are two classes of systems, namely, man-made and natural systems. We will be looking at both types of systems in this chapter.

Man-made systems

These systems involve everyday activities and are necessary to sustain ordinary life on a day-to-day basis. A partial list of such systems is provided as follows.

- The food growing and distribution system
- The e-mail system
- The telephone system
- The gas distribution system
- The drug distribution system
- The transportation system
- The automobile sales and repair system
- The department store system
- The hardware distribution system
- The school system
- The banking system
- Business information systems

    An enormous amount of effort is devoted to building, operating, and maintaining these systems, and by and large, we put a lot of person hours into them. They are generally not scientific, but meat-and-potatoes systems that we need every day. But to keep them up and running efficiently, a lot of thought goes into them every day.

Natural systems

Many folks have been exploring and thinking about natural systems, and for a long time, and to our great benefit. Here are a dozen examples:

- Copernicus thought about the solar system
- Descartes thought about logic and metaphysics systems
- Galileo thought about the solar system and pendulum
- Maxwell thought about the electromagnetic system
- Salk thought about the immune system
- Crick and Watson thought about the biologic system
- Darwin thought about the hereditary and evolution system
- Einstein thought about the gravitation system (as did Newton)
- Franklin thought about the optics and electricity systems
- Rachel Carson thought about the ecological system
- Freud thought about the mental system in the humans
- Da Vinci thought about art and warfare systems

In any of these cases, can you begin to imagine how these thinkers did what they did? Can you imagine the train of thought they were on? Can you find a place to start in terms of the first questions you would ask?

By and large, the earlier cited systems are natural, meaning that they exist in nature. And we constantly devote our best talent and thinkers to trying to gain a deeper understanding of these natural systems.

A "systems" perspective

When we are trying to define and solve a problem, we attempt, under this perspective, to take a broad view. We try to see the problem in its larger context, as part of a system. We prefer to keep away from narrow interpretations of the problem. Why? Since it has been shown that we can be, and usually are, more successful with the broader "systems" perspective. In other words, we do it because it works. An illustrative example will help to make the point.

Some years ago, Barry Boehm, a leader in software engineering, documented his experience with building a system at TRW [2]. The system was supposed to have a response time of 1 s and cost some $30 million. As it turned out, the team was not able to meet that response time at that cost. After some analysis, they decided they could meet the requirement, but at a cost of $100 million. The team apparently focused on the matter of "how do we meet that requirement, and at what cost?" That was "the problem." But with some internal advice, they went to their customer and set forth what they had learned, which was

- Can only meet the 1 s response time at a cost of $100 million
- Can meet a 4 s response time at the original cost of $30 million

So, looking at a broader perspective, they viewed the customer as part of the "system," and they also introduced a wider view of the issue at hand. Both of these were in consonance with a "systems" perspective—and to the credit of the project team. So rather than being bogged down in trying to obtain a 1 s response time, the broader view gave them an opportunity to be successful. Oh, by the way, the customer chose the 4 s response time solution at the original project cost of $30 million.

The earlier example illustrates what it means to take a broader "systems" view. It is likely to pay immediate dividends in terms of practical problem-solving. Similar approaches are suggested when we look at the following "descriptors":

- Holistic
- Integrated
- Generalized
- System-wide
- Fusion
- Top-level

These are to be contrasted with a non-systems view with these notions:

- Limited
- Singular
- Differentiated
- Unilateral
- Pointed
- Selected

From this point, we move on to concepts having to do with the "systems approach."

The systems approach

The general approach in dealing with man-made systems is called the "systems approach." Some of the thinking that has gone into clarifying what this approach means is described as follows.

An early purveyor of the "systems approach" was C. West Churchman, who wrote a seminal treatise on the matter [3]. His approach, it seems to this author, is cautious. He seems at times to be put off by the notion that he is advocating or "selling" the systems approach. But in the final analysis, he comes down squarely in favor of it.

In the context of Prof. Churchman's book, we need to keep in mind that his systems approach is essentially a problem-solving method. To address and solve many "societal" problems, our preferred approach is from a systems point of view. Well said, and well accepted. Another element in Churchman's approach is that of a fabulist. He looks at the well-known story of the blind men that examined a part of an elephant, each coming to a conclusion based upon this limited contact. Each conclusion is wrong, since, of course, they missed the totality of the touches—a clear example from the world of fables.

Another quite different approach is taken by this author, considering instead the task of building a large-scale system. The focus tends to address the question:

- In the systems approach, what exactly are the kinds of things one does in order to develop a sound and successful system? [1,4]

A brief citation of the answers to that question can be found later for the following 12 notions.

1. **Establish and follow a systematic and repeatable process**
   Process is very important as part of the systems approach, and so we make sure that we have one that is well defined and can be used by different people.
2. **Assure interoperability and harmonious system operation**
   The various components of a system must be able to operate together, respecting the need for harmony and interdependence.
3. **Consider alternatives at the various steps of design**
   This is a critical part of the approach. Here we broaden our view and look at a variety of possibilities, choosing the best one among several.
4. **Use iterations to refine and converge**
   We distinctly do not expect that we will do it right the first time. We deliberately use iteration to our advantage and as a means of converging to the correct solution.
5. **Create a robust and slow-die system**
   This is a part of the design rules for our system: few if any single-point failures. Build in lots of redundancy to have the system degrade gracefully.
6. **Satisfy all agreed-upon user/customer requirements**
   In early stages, we treat requirements as subject to trade-offs by exploring with user/customer base. Once these requirements become firm, they are agreed upon and not subject to change.
7. **Provide a cost-effective solution**

The overall perspective is based upon cost-effectiveness considerations. We select the most cost-effective solution to the stated problem and deal directly with the matter of how to measure effectiveness.

8. **Assure system sustainability**

    Even in today's world of abundance, we require that the system be self-sustaining as much as possible. We treat sustainability as an "ility" and do the calculations rather than just assume we can build such a system.

9. **Use advanced technology, at appropriate levels of risk**

    Technology is very important in most of our systems. It must be part of our systems approach, and therefore, we look explicitly at what choices we have and make a conscious decision to use certain technology and not others, on the basis of risk.

10. **Consider all stakeholders and their concerns about the system**

    This broadens item number six mentioned earlier and makes sure that (1) we understand all the stakeholders and (2) have taken their often disparate concerns into account. We do not necessarily expect to satisfy all their points of view simultaneously, recognizing they may be in conflict.

11. **Design and architect for system integration**

    Efficient and effective system integration is the name of the game in building large-scale systems. If we design for integration, we are more likely to be successful. This is one of the greatest challenges in the systems approach.

12. **Employ systems thinking**

    "Systems thinking" is the bedrock of the systems approach. We accept that premise and move on to a deeper understanding of what this is and how to do it.

Systems thinking

Certainly a touchstone for exploring "systems thinking" has to be Senge's classic work on *The Fifth Discipline* [5]. The basic rationale is that a goal is for enterprises to become a learning organization. This is not a high bar; what organization would not aspire to such a goal? So Senge moves on to say that to meet such a goal, there are five disciplines upon which to focus, namely

1. Personal mastery
2. Mental models
3. Team learning
4. Building shared vision, and
5. Systems thinking

Personal Mastery has to do with developing and maintaining skill sets at the personal individual level within the organization. **Mental Models** are views of the world and of the organization that will tend to become views within the organization. **Team Learning** is an attribute of the teams within the organization, considered to be crucial in terms of how the enterprise functions. **Shared Vision** deals with building a vision within the organization that most people buy into. Senge illustrates this by pointing to IBM's "service," Polaroid's instant photography, Ford's public transportation, and Apple's computing power for the masses. Finally, he comes to "systems thinking."

According to Senge, **Systems Thinking, as the fifth discipline, integrates the disciplines into a coherent body of theory and practice.** This is his definition and use, and his entire exposition is based upon this notion. In his primary book, he gives many examples of this type and form of systems thinking. One might say that this is his contribution to the idea of systems thinking as well as the various implementations thereof. It is not the only way that systems thinking is dealt with by a variety of investigators. However, it is one of the most widely accepted.

In 2015, Arnold and Wade published an overview and seminal account of systems thinking [6]. They first addressed the matter of a "system test," a set of criteria by which one decides whether or not we have a true definition of systems thinking. That test has three parts:

a. A function, purpose, or goal
b. A set of elements, and
c. Interconnections

The notion here is that systems thinking must have all three elements to complete (a necessary condition).

Arnold and Wade then proceed to look at many definitions in the literature in the light of this system test. The first of these is the work of Barry Richmond [7]. The claim here is that Richmond was the first to coin the phrase "systems thinking" back in 1987. He led the charge with the comment that interdependency demands systems thinking.

Other notions expressed by Arnold and Wade include the following:

1. A reductionist approach is generally considered a nonsystems-thinking approach.
2. Senge's definition is somewhat vague, and does not pass the "system test".
3. Dynamic complexity can be associated with systems thinking.
4. A comparison of eight approaches to the definition of systems thinking.

Rechtin's heuristics

Eberhardt Rechtin was surely one of our master engineers, having held several key management positions and also having written an important book on the matter of architecting large-scale systems [8]. Among other seminal ideas, he lists a set of heuristics for the systems. In effect, this list can be considered ideas or thoughts about systems. They could have been called "thinking about systems." So they definitely deserve to be a part of any chapter on "systems" and on "thinking." Here are ten selected heuristics areas that the systems engineer needs to keep in mind:

Selected Heuristics from Eberhardt Rechtin [8]

Location	Topic
Chapter 1	Simple and correct solutions
Chapter 2	Leverage and interfaces
Chapter 2	Expansion of a difficult problem
Chapter 2	Software mistakes
Chapter 3	Multiple directions
Chapter 3	No all party optimum
Chapter 3	Models and reality
Chapter 3	Accept analogies
Chapter 4	Take time to reflect
Chapter 9	Systems and countersystems

The reader is encouraged to go beyond the earlier sample to the full richness of Rechtin's heuristics list in the appendix of his book.

Thinking in systems

Donella (Dana) Meadows [9] is a name almost always in evidence when discussing thinking in systems. She passed away in 2001, but was an integral part of new understandings of systems and how they work, and not work. She and her husband were important players in *The Limits to Growth* book and idea. Much of her work was connected to the MIT System Dynamics Group, and also to Jay Forrester who was the founder of the group. Thus, a large part of her approach was related to Jay Forrester and his modeling creations. This will be discussed in somewhat greater detail later in this chapter.

Meadows's book [9] is a touchstone for her thoughts about systems and about thinking, and about modeling of these systems. A singular part of the quantitative theory is that systems exhibit feedback which, depending upon how that is achieved, can make the "system" stable or unstable. We normally expect stability, as in a servomechanism. But Forrester's models contain other elements that promote stability and yield definitive as well as quantitative answers.

Here are some of the other points that are made by Meadows that bear upon the matter of thinking in systems:

1. A system is more than the sum of its parts
2. Systems have interconnectedness and a corresponding flow of information
3. Three important features of systems are resilience, self-organization, and/or hierarchy
4. There are always limits to resilience
5. Often, real systems exhibit nonlinear vs. linear behavior
6. Leverage points in systems are often counterintuitive

Each of these points, and others, deserve a fair amount of thinking.

Forrester

Jay Forrester is the genius who invented a most serious way of analyzing a system [10,11]. That system has been called by many names, including "System Dynamics," "Industrial Dynamics," "Urban Dynamics," and "World Dynamics." It is the basis for many researchers' methods that have used the system to analyze the world and its numerous systems. It has been instantiated in the software language known as "DYNAMO." Forrester's work has turned out to be the basis for many "systems" thinkers, especially with respect to quantitative analysis of complex systems.

Monat and Gannon

These authors [12] make interesting and useful points regarding the use of systems thinking. They are particularly focused on "real-world" problems and issues. They use examples to make their case with such ranging into the worlds of addiction, measles, and ISIS (the terrorist group). Here is their general perspective regarding systems thinking, followed by an important observation:

- Systems Thinking "focuses on the relationships among system components and the interactions

of the system with its environment, as opposed
to focusing on the components themselves. It is
holistic (integrative) thinking instead of analytic
(dissective) thinking"
- Systems are also dynamic and are constantly subject to forces and feedback mechanisms

It is to be noted that these authors lean heavily upon the work of Jay
Forrester and his breakthrough modeling in the system dynamics world
[10,11]. They see systems thinking as including both analytic (reductionist)
thinking as well as statistical thinking. Some investigators are not as
inclusive, rejecting reductionist approaches as not consistent with systems
thinking. Appropriately, they pay special attention to the use of tools as an
integral part of solving real-world problems. Such tools include

- The iceberg model
- Causal loop diagrams and feedback
- Behavior-over-time graphs
- Stock-and-flow diagrams
- Dynamic modeling
- Archetypes
- Systemic root cause analysis

It is clearly useful to have a set of such tools available to work on real-
world problems and issues.

Systems thinking and complexity

An important researcher [13], and erstwhile colleague of Russell Ackoff,
wrote a book about systems thinking and complexity. Although the book
ranges far and wide on a variety of system-related matters, one of his
main themes is that

- It is necessary to simplify complexity

So we are building complex systems, and we are also confronted with
complex systems, and the task at hand, to cope appropriately, is to **be able
to simplify**. Part of the answer lies in holistic (systems) thinking, which
involves three foundations:

1. Self-organization
2. Operational thinking, and
3. Iterative design.

Interactive design plays an important role in this author's scheme. It is
designing the future and inventing ways to bring it about. We note the

fact that iterations are part of the "systems approach" as articulated earlier in this chapter. There are no negatives regarding the notion of iteration. In fact, it is part and parcel of the process of building systems. We use it deliberately to converge to the answer we are looking for.

Another theme in this author's approach is the notion of designing inquiring systems. Such systems are allowing users to move forward by asking questions—by inquiring. When one is asking and answering questions one is also doing a lot of thinking.

So here is a short form citation of some of the "systems" aspects that this author deals with

1. The imperatives of interdependence
2. The need to reduce endless complexities
3. The need to simplify

If we are not up to several of the earlier, we may wind up not able "to focus on the relevant issues and avoid the endless search for more details while drowning in proliferating useless information." We note that Rechtin [8] makes a similar point in his list of heuristics. Rechtin's claim in this regard is that "amid a wash of paper, a **small number of documents** become critical pivots around which every project's management revolves." Let's keep that one in mind. It's an important point and too many systems engineering teams are producing a lot of paper that is not really useful in terms of building and managing new systems.

Creative holism

Michael C. Jackson has embraced creative holism for managers and featured it in his wonderful book on "systems thinking" [14].

Jackson is well known for his "applied systems thinking" with its articulation of "hard" and "soft" systems thinking. The former is supported by classic quantitative methods.

Some of the approaches in this category of thinking include system dynamics (one more time, and with justification), organizational cybernetics, and complexity theory. But, as was set forth by Jackson and others, hard system thinking has its limitations, opening the door for other approaches. A prominent such approach is "soft" systems thinking. Jackson appears to leave that ground to Peter Checkland, and so we move on to a brief discussion of his work and perspective.

Soft systems thinking

Apparently, some of the limitations of hard systems thinking led to the formulation of soft systems thinking. Checkland [15] was a key player in that formulation and in the further explanation of its meaning. Here is one definition of what he has called a soft systems methodology:

- Soft systems methodology is for tackling real-world problems in which known-to-be-desirable ends cannot be taken as given. Soft systems methodology is based upon a phenomenological stance

Sitting alongside of this theory is the notion, and body of knowledge, known as appreciative systems theory. This has gained some traction and is a theory unto itself at this time. The reader is urged to explore Checkland's work and writings in terms of its far-ranging scope and explanations of this type of analysis and thinking.

Systems thinking and learning

Yet another view of systems thinking is presented by Haines [16], with the following description of the subject at hand:

- Systems Thinking is "a new way to view and mentally frame what we see in the world; a worldview and way of thinking whereby we see the entity or unit first as a whole, with its fit and relationship to its environment as primary concerns"

Haines sees the progression to systems thinking in terms of moving from "chaos and complexity to elegant simplicity." This is a common theme that we have seen from several investigators. Haines also contrasts systems thinking with "machine age" thinking, the latter being driven by reductionism and analysis. His arguments lean a lot upon looking at living systems; systems that occur in nature. His framework for thinking leads to seeing relationships and patterns, which in turn lead to better problem-solving.

Thinking in loops

Systems thinking, as per O'Connor and McDermott [17], is "thinking in loops rather than in straight lines." At times, these are feedback loops, as these authors have subscribed to the viewpoint of Jay Forrester. They accept the notion that there are two types of feedback loops, reinforcing and balancing. They also believe that there are great benefits through systems thinking. Their view of a system is simply that it's an entity "that maintains its existence and functions as a whole through the interaction of its parts."

Laws and laws and laws

This author [18] is an incredible thinker along with his extremely high productivity and range of investigations. In this particular treatise, reprinted more than 20 times over a period of a quarter of a century, the author emphasizes systems thinking in the form of "laws" that are articulated by the author and that apply to systems in general. Here is a sampling of some of the laws that are cited, without further explanation of what they mean:

- The law of conservation of laws
- The law of happy particularities
- The law of medium numbers
- The generalized law of complementarity
- The axiom of experience
- The banana principle
- The brain-eye law
- The composition law
- The diachronic principle
- The perfect systems law
- The principle of difference
- The principle of invariance
- The square law of computation

Clearly, this author's thinking ranges far and wide.

This author also cites what he calls the **Systems Triumvirate**, three key questions that define, for him, the essence of systems thinking, namely:

a. Why do I see what I see?
b. Why do things stay the same?
c. Why do things change?

It should also be noted that this author has produced nearly 40 books that pertain to the software development life cycle. This is a most notable achievement, contributing to a world in which there is much depth as well as variety in terms of getting new systems up and running.

An illustrative example—system architecting

It is now time to look at an example that relates directly to the notions of this chapter in the light of what the systems engineer does as a key activity. That activity is the architecting of systems. For this author, there is nothing more important than "systems architecting" as a representation of what the systems engineer is responsible for.

DoD Architectural Framework

The Department of Defense (DoD) took the lead in addressing the matter of system architecting by defining a "Framework" for doing so [19]. This framework started with the definition of "views" of an architecture, of which there were three, defined as

- The Operational View
- The Systems View
- The Technical View

It was not clear as to exactly how to construct an architecture from these views, in spite of the fact that several steps were set forth to do so. However, the DoD made it clear that these views could (and should) be expanded. That led to a rather larger list of views that provided more and more detail as to the nature and content of such views. This expansion led to the articulation of essential and supporting views, with the former being:

AV-1: Overview and Summary Information (an "all" view)
AV-2: Integrated Dictionary (another "all" view)
OV-1: High-Level Operational Graphic Concept
OV-2: Operational Node Connectivity Description
OV-3: Operational Information Exchange Matrix
SV-1: System Interface Description
TV-1: Technical Architecture Profile

There are a total of 18 supporting views, 13 of which are "systems" views. The point of this presentation is to look at the basic approach, which is clearly a "**drill-down**" procedure. So we have to conclude that **the DoD Architectural Framework (DoDAF) approach is basically a reductionist one**. Thus, this is the opposite of the so-called "systems" approach. This is an observation, not a judgment, and it certainly has its place in this and many other problem-solving settings. From here, now let us look at another, and quite different, architecting approach.

The EAM architecting approach

In contrast to the earlier DoDAF procedure, we cite now an alternative approach to the architecting of systems. This is the Eisner's Architecting Method (EAM)approach [4], which is distinctly different. The basic four steps of this approach are as follows:

1. Functional decomposition of the system
2. Synthesis
3. Analysis, and
4. Cost-effectiveness evaluation

Functional decomposition   The very first step in this procedure is to decompose the overall system into its functions and sub-functions. We will architect the system, in the main, at the sub-function level. This is basically a **reductionist step** in that we are looking at breaking the system down (by function) into the "tasks" that the system is to perform. This sets the stage for the next, and crucial, step of synthesis.

Synthesis   This step involves producing a chart of sub-functions (rows) vs. design approaches (columns) for several (nominally, at least three). So we are looking at multiple architectures (at least three) in this process. These are "**alternatives**" and are critical to the overall approach. The alternatives represent low-cost, high-effectiveness, and best value (knee-of-the-curve) architectures. Additional alternatives may be defined if the architecture team wishes to do so. By definition, each column in this chart represents an architecture so that we have a clear and simplified version of an architecture. So there are two observations that we wish to make about this procedure and example:

1. By defining at least three alternatives, we are following a "systems" approach rather than a reductionist approach, and
2. We are attempting to simplify a complex problem, also in consonance with a "systems" approach

The bottom line is that this example mostly demonstrates a "systems" approach, but not without some element of reductionism in the functional decomposition step.

Analysis   If we have at least three alternative architectures from the previous procedure, the next step is to analyze the three to see which appears to be the preferred architecture. There are several possibilities in terms of executing this step, falling in the category of evaluating alternatives. A weighting and rating scheme is used in the EAM, and an effectiveness score is calculated. Also, the cost is determined using standard cost-estimating procedures. At the conclusion of this step, we have (at least) three architectures, and three corresponding estimates of their costs and effectiveness values.

Cost-effectiveness evaluation   The final choice is then made as to a cost-effective solution, from the above numbers as well as any ancillary factors.

A final overview of the EAM approach to systems architecting shows that there is a distinct selection of the "systems" procedure in the use of alternatives. However, there are definitive signs of the reductionist approach in the functional decomposition as well as the analysis step. Thus, it is possible to have a mixed (hybrid) approach, perhaps dominated by the holistic view provided by alternatives and the subordinate view gained by "drilling down."

Closing suggestions for the systems engineer

We have been exploring various notions regarding systems thinking and systems in this chapter. Now the question is— **how can these be useful to the systems engineer**? As we close the chapter, we try to be specific about the answer to that question, with the following list of suggestions.

1. As per Rechtin, **simplify your design** as much as possible, while still building in redundancy to construct a "slow-die" system. Remember his KISS (Keep It Simple, Stu..d) principle and his treasure trove of heuristics
2. Try using the **EAM approach for system architecting**, which by its nature is simple and compelling (there are three architectures on one sheet of paper)
3. Pay special attention to the details of your **functional decomposition**. Decompose to three levels only (where level one is the system name), and architect at that level
4. Consider designing for **both low cost and best value** as you compete in the world of building large-scale systems
5. Make sure to **formulate alternatives** for all the important parts of your design
6. Use **systems thinking in your measurements** programs; measure just what you need, no more
7. Consistently use **cost-effectiveness as your approach** to your customer. Find a cost-effective solution and prove that it's the best answer for him or her
8. As you design your system, practice asking what you consider to be the **"right" sequence of questions**. Address these questions one at a time, and with the idea that you will be able to look up, down, and sideways for solutions
9. Design to **facilitate integration** of the parts of your system
10. Embrace the **"systems approach"** and **"systems thinking"** as much as possible, and remember that such a commitment involves the "thinking-up" perspective

References

1. Eisner, H., *"Topics in Systems"*, Mercury Learning and Education, 2013.
2. Boehm, B., "Unifying Software and Systems Engineering", *Computer Magazine*, March 2000, pp. 114–116.
3. Churchman, C. W., *"The Systems Approach"*, Dell Publishing Company, 1968.
4. Eisner, H., *"Systems Engineering: Building Successful Systems"*, Morgan & Claypool, 2011.
5. Senge, P., *"The Fifth Discipline: The Art and Practice of the Learning Organization"*, Doubleday, 1990.

6. Arnold, R. and J. Wade, "A Definition of Systems Thinking: A Systems Approach", *Procedia Computer Science*, Vol. 44, 2015, pp. 669–678.
7. Richmond, B., "Systems Dynamics/Systems Thinking: Let's Just Get on with It", *International Systems Dynamics Conference*, Sterling, Scotland, 1994.
8. Rechtin, E., *"Systems Architecting: Creating & Building Complex Systems"*, Prentice-Hall, 1991.
9. Meadows, D., *"Thinking in Systems"*, Sustainability Institute, Chelsea Green Publishing, 2008.
10. Forrester, J., *"Principles of Systems"*, Pegasus Communications, 1990.
11. Forrester, J., *"Urban Dynamics"*, Pegasus Communications, 1969.
12. Monat, J. and T. Gannon, *"Using Systems Thinking to Solve Real-World Problems"*, Systems Engineering Program, Worcester Polytechnic Institute, 2017.
13. Gharajedaghi, J., *"Systems Thinking"*, Second Edition, Elsevier, 2006.
14. Jackson, M. C., *"Systems Thinking: Creative Holism for Managers"*, John Wiley, 2003.
15. Checkland, P., *"Systems Thinking, Systems Practice"*, John Wiley, 1999.
16. Haines, S., *"Systems Thinking & Learning"*, HRD Press, 1998.
17. O'Connor, J. and I. McDermott, *"The Art of Systems Thinking"*, HarperCollins, 1997.
18. Weinberg, G., *"An Introduction to General Systems Thinking"*, Silver Anniversary Edition, Dorset House, 2001.
19. The DoDAF, Department of Defense, dodcio.defense.gov/Library.

Exercises

3.1 Identify eight "systems" not cited in this chapter.

3.2 How would you define "the systems approach" (not as in this book)?

3.3 What is your best overview definition of "systems thinking"?

3.4 What are your five elements of (components of) "systems thinking"?

3.5 What are five key elements in the process of architecting systems?

chapter four

Other thinking perspectives

Introduction

The previous chapters addressed a variety of ways of thinking. However, these were just the "tip of the iceberg" in terms of the size and scope of thinking ideas and patterns. This chapter moves forward with several other ways of approaching the matter of "how to think."

Reductionism

An important person with a special perspective on thinking was **René Descartes** (1596–1650), who declared, "I think, therefore I am" (*cogito ergo sum*). Descartes was a philosopher, scientist, and mathematician with many early contributions, and this existential statement was and remains rather startling. And somehow, it has led to the thinking that might be called a form of reductionism. That is, it has supported the notion that we can solve large-scale problems by first breaking them into smaller problems. Then we can systematically solve these smaller problems and put the pieces together to solve the overall problem. This is a logical approach, but it's not necessarily workable all the time. This author was involved in at least one such project where it did work. Here's what happened.

The Department of Transportation (DOT) took on the problem of trying to find out what the relationship might be between the existence of a fleet of supersonic transport aircraft and the possible increase in cancer. This was called CIAP, the Climatic Impact Assessment Program, and was run out of the Office of the Secretary in the DOT. This "big" problem was broken into subordinate smaller problems, approximately, as follows:

1. Developing a scenario of fleets of supersonic transport aircraft in operation
2. Estimating the effluents that would be emitted from these aircraft
3. Evaluating the effects of these effluents on the atmosphere
4. Calculating the special influences on the ozone layer (i.e., depletion)
5. Estimating possible increases in cancer as a result of the previous item (4)

These five subordinate problems and their solutions were then considered, in tandem, to "solve" the overall problem. This author remembers this process as a quite well-conceived and executed problem in which reductionism worked, and did so very well over a period of about 3 years. The difficulty was making sure that the overall study design was carefully developed and implemented. Another word that is sometimes used for this type of thinking is simply "decomposition." The sequence is as follows: decompose—solve each piece—synthesize.

Breakthrough thinking

Some time ago, the Harvard Business Review put together a list of articles pertaining to the matter of how to do "breakthrough thinking" [1]. Here's a brief overview of these articles, all of which have something to say about thinking in the business world.

Killing creativity [2]

It is claimed that managers may need to change their ways of thinking to achieve the increases they seek in productivity, control, creativity, and efficiency. It has to do with intrinsic motivation and how to achieve it. Managers need to focus on the three elements of such motivation— expertise, thinking flexibly, and imaginatively.

Empathic design [3]

These authors define empathic design that will seriously improve products and the relationship between developers and customers. The steps in empathic design are as follows:

1. Stronger observation
2. Capturing data
3. Reflection and analysis
4. Brainstorming for new solutions
5. Developing prototypes for new possible solutions

A working whole brain

This perspective involves getting the "whole brain" to function in an enterprise and getting collaborators to work together [4]. Here are some suggestions for action:

1. Manage the creative process
2. As a Chief Executive Officer (CEO), look for cooperation from people unlike you

3. Depersonalize conflict
4. Understand personality types, as per the MBTI (Myers-Briggs) and the Hermann Brain Dominance Instrument

Managing creativity [5]

Can creativity be managed? These authors say "yes" to that question and attempt to show how. This is approached from the perspective of a film director who has to blend the activities of a group of people. This director was a well-known person, and the authors studied his approach in great detail. If you know what you're doing and have a lot of experience making movies and handling people (gently), then it is absolutely possible to "manage creativity."

Stifling creativity [6]

The opposite side of the coin, in relation to the earlier, is "stifling" creativity. The CEO of this fruit juice company was trying to understand how it was that creativity was losing ground. It was reflected in the stagnant growth and the attitudes of many of the key employees. Conclusion? Too stubborn about enforcing a culture of tradition, self-discipline, and respect for authority. It was time to change that and encourage a more open and creative behavior. You have to think long and hard about how to do that.

Innovation

One of our great thinkers in the management field gave us some insight into this problem and issue. That was Peter Drucker [7], never at a loss in terms of what might be going wrong and how to "fix" the problem. Opportunities for innovation can come from inside or outside the enterprise. Rather than a flash (or two) of genius, Drucker prefers, and recommends, a much more disciplined approach, looking within and outside the company for such features as incongruities, changes in demographics, market changes, and new knowledge.

Professor Drucker was a very special contributor to a better understanding of the details and complications of the fine art of management. He was one of our great thinkers.

Interpretive management [8]

What does "interpretive management" mean? These authors suggest that more conventional management approaches be replaced by those used by design managers. They need to look for ways to promote conversations about the future. This change requires adjusting the ways of thinking as well as behaving.

The logic of high growth [9]

These authors pose an important question—"why are some companies able to sustain high growth and profits, and others are not"? They studied 30 companies around the world and found that the key to an answer had to do with assumptions about strategy. They had broader views of what was constrained and what was not. They replaced conventional logic with what they called value innovation.

Other aspects of breakthrough thinking

Nadler and Hibino [10] strongly suggested that we need to change the way we think and therefore how we solve problems. They say that there's a better way, elucidated by seven principles, three of which are as follows:

1. The uniqueness principle (each problem is unique)
2. The purposes principle (strip away nonessential aspects of the problem)
3. The solution-after-next principle (work backward from an ideal solution)

The reader is referred to the reference for a more complete treatment.

So what then is a reasonable definition of "breakthrough thinking"? This author offers the following perspective on this set of words and concepts:

- **Breakthrough Thinking** is literally "thinking outside the box" that moves beyond conventional approaches and promises, as well as achieves better results within appropriate levels of effort and commitment. It is thinking well beyond the norms of the day and suggests various forms of innovation.

Lateral thinking

Lateral thinking was conceived of and explored in some detail by Edward de Bono, a well-known psychiatrist and thinker from the United Kingdom [11]. The basic idea is that when your thinking gets "stuck," a good alternative is to move "laterally" from where you are to some other place. A frequent mistake is to dig more deeply where you are and that gets you into more and more trouble. Here are a few of de Bono's contrasts between vertical and lateral thinking:

- Vertical thinking is selective, while lateral thinking is generative.
- Vertical thinking moves only if there is a direction in which to move; lateral thinking moves to generate a direction.
- Vertical thinking is analytical, while lateral thinking is provocative.

Let us now look at a few examples of "lateral thinking." In the first example, a company is thinking about their strategic plan. They build and install airport surveillance radars, largely for the Federal Aviation Administration (FAA). They are dominant in the field but do have worthy competitors. One path for them to take is to try to penetrate more deeply and take a larger share of that market. Yet another path is to look sideways (laterally) at other potential markets for that radar and variants thereof. This lateral look opens up other possibilities such as harbor radars, upgrading to air route surveillance radars, and even a deeper look at foreign markets. A serious lateral look threatens deeper penetration of the market they are in. What to do? They need to struggle with the details of this issue, but lateral thinking puts them in a position where they can consider other sensible alternatives. This is an important part of thinking about their strategic directions for the future.

In a second example, a company's executive board of four people is digging deeply into a possible acquisition. They are gathering and analyzing more and more data. As a result, they're not ready to make a decision. Meanwhile, an impatient fifth member of the board calls a friend for lunch. That friend suggests that he's ready to talk acquisition, and within the week, there's a meeting to look at his data. An "intent to acquire" document is signed and the lateral alternative is chosen, without further fanfare and analysis.

A third illustrative example involves the interaction between two competing countries. One country is dominated by military thinking and they are looking at reasons to do a preemptive war strike. The competing country is studying how to start a serious negotiation. Too late—the first country implements a strike, choosing a lose–lose "solution" instead of a win–win alternative. Could this happen? It's happened before and likely to happen again when vertical thinking can dominate.

Thinking hats

De Bono continued his pursuit of ways of thinking by conceiving of the notion of "thinking hats" [12]. This involved establishing a group of interacting people in which each person had a role and point of view reflecting different perspectives. Here are the various thinking hats:

1. White Hat: Facts and Figures
2. Red Hat: Emotions and Feelings
3. Black Hat: Cautious and Careful

4. Yellow Hat: Speculative and Positive
5. Green Hat: Creative and Thinking
6. Blue Hat: Control of Thinking

This six thinking hats approach demonstrates de Bono's creativity and represented a new way of exploring group behavior. The participants are all aware of their roles and have to figure out how to behave. They have to give the other people "room" to adapt to these roles and have to listen very carefully to each other. Staying in role is very important to the overall functioning of the group. The blue hat person is extremely important and is set up to summarize along the following lines:

1. What has been achieved?
2. What specific outcomes have been produced?
3. What conclusions have been reached?
4. What solutions have been found?
5. What steps are suggested next?

This is certainly a most interesting perspective regarding the ways of thinking in a group setting. And de Bono is making an important contribution as he introduces new approaches to thinking about an important problem—that of human behavior in a group setting.

Dr. de Bono has also paid special attention to how his approach might influence and improve the various behaviors in management. His book on the subject [13] cites various examples of how management can use the principles of lateral thinking. His *Super Mind Pack* sets forth some 14 "games" that can be played to explore what he calls "thinking for action" [14].

Design thinking

A considerable amount of effort has been devoted, in the business world, to the notion of "design thinking." A leader in this field has attempted to show "how design thinking transforms organizations and inspires innovation" [15]. Design thinking focuses on innovation in the marketplace, three features of which are inspiration, ideation, and implementation. This embodies a thinking shift, from problem to project, implemented by smart teams. These teams are trained, as part of the process, to avoid any aspect of groupthink (see also chapter one). These teams also work explicitly with clients to come up with new and exciting design approaches. These types of design approaches also involve the notion of "integrative" design. This fits rather well with the "systems" ideas set forth in the chapter on systems.

An important aspect of design thinking, according to Tim Brown, is that of "prototyping." This feature moves the design team beyond the "thinking" phase and into an activity in which there is something to see, touch, and feel. Emphasis is placed upon exploring options, which brings us face-to-face with having sets of alternatives rather than just one design to deal with. We recognize the formulation of alternatives as a part of "systems thinking" and systems engineering.

Special attention is paid to "human-centered" design, recognizing that the human is a natural part of our systems, as user (operator) or maintainer, or both. This is an expansion of what is sometimes called "human factors."

"Storytelling" is also encouraged, even though it might be called "rambling" in another team's context. This pushes in the direction of "thinking outside the box." So is "visual thinking," supported by a certain amount of picture and sketch drawing.

No new idea is rejected no matter its source. It can come from today's newspaper, a TV program, a commercial, or from a younger sibling of a member of the team. If the idea has merit, it is brought into the process and the fray. Even "disruptive" ideas are encouraged, as per the next subject explored here.

Disruptive thinking

Part of the world of thinking has now been instantiated in the term "disruptive" thinking. Some documentation of this field and concept is provided for us in the book *Disrupt* [16]. In this treatise, Luke Williams claims that he and his team can think what no one else is thinking and go on to do what no one else is doing. It is disruptive in that it turns consumer expectations "upside down" and brings a whole industry into its next generation.

Apparently, there are some five stages of disruptive thinking suggested by Williams:

1. Develop a disruptive hypothesis
2. Define a disruptive market opportunity
3. Develop several disruptive ideas
4. Shape them into a single disruptive solution, and
5. Present a disruptive "pitch" to persuade all potential stakeholders to invest or adopt

So here we see a specific set of steps to follow, which has a good chance of leading to a powerful market disruption.

Here are a few examples suggested by Williams:

1. Socks in sets of three, none matching
2. Having plumbing, vents, and electrical outlets on the outside of buildings, per the Pompidou Center in Paris
3. The Zipcar

One is moving from hypothesis to opportunity, in general, as in the example later:

- Hypothesis—what if mops did not use water
- Insight—a common failure is too much water
- Opportunity—we can clean floors without using water, which is not a solution, but rather an opportunity

The full sequence therefore appears to be: hypothesis to observations to insights to opportunities to new ideas. In moving through this sequence, it is important to involve end users in the process to test and review, each step of the way. As with design thinking, it is useful to build and test prototypes to be able to see and touch real hardware and software, in a "quick and dirty" manner. Finally, Luke Williams reminds us that the term "disruptive" likely was borrowed from Clayton Christensen from his book on the "Innovator's Dilemma" [17] in reference to "disruptive technology."

Ackoff's problem-solving

Russell Ackoff [18] has been one of the gurus of management and problem-solving (see also chapter three). His type of problem-solving has been special and unique, involving thinking deeply and using fables to make his points. He has also addressed the whole matter as the "art of problem solving." Here are some aspects of Ackoff's approach to problem-solving:

1. Look at desired outcomes (objectives)
2. Investigate possible courses of action
3. Look at the overall environment
4. Articulate the relationship between all relevant variables
5. Identify constraints that may be part of the overall problem

Dr. Ackoff, in his "classic writings" [19], has also relied on systems concepts and approaches. He has been broad in his thinking but also deep in his understanding of management and the human condition. He is credited with defining the "five C" approach to considering the issues of management, namely

1. Concern
2. Competence
3. Communication
4. Courage
5. Creativity

Yes—it's not possible to be creative without a lot of attention to the "thinking" element.

Critical thinking

Diane Halpern had been dealing with critical thinking for a long time and became a champion of what it is and how it's done [20]. Her fifth edition book pushes the envelope and clarifies some earlier points in her investigations. In dealing with critical thinking, she examines the roles of such topics such as

1. Reasoning
2. The use of memory
3. Hypothesis testing
4. Problem-solving
5. Creativity
6. Language
7. How to make improvements
8. How thinking really works

Of particular importance is her perspective on analyzing arguments. In a sense, that's what critical thinking is all about. It is looking at various arguments that have been set forth and being able to analyze them in great detail, if and when necessary. She also uses diagrams to help explain and explore these matters. This supports some of the ideas in this book regarding "pictures" to help elucidate thinking.

Creativity and ideas

So an immediate question might be—what is it that makes a person creative? A simple answer is likely to be: he or she has been "trained" or is so inclined to naturally think about the world in new and different ways. These are ways that are uncommon and lead to ideas that are foreign to most other people. So we are in the world of extraordinary thoughts and ideas. And in that world, it's like mastering many things—it takes practice, practice, practice.

A perspective regarding creativity was set forth by a professor at the University of Chicago [21] after interviewing some 90 people on this subject. These people appeared to be especially creative in their approach to life and problem-solving. He was led to searching for two traits:

1. Curiosity and openness to problems and situations and
2. An almost obsessive perseverance

In this author's experience, many successful people point to the latter (perseverance) as the most important attribute in their lives. Apparently, just staying with it, and working the problem (obsessively) go a long way to being creative as well as successful. As a data point, I am reminded of a notable incident. When Robert Pirsig was trying to find a publisher for his book *Zen and the Art of Motorcycle Maintenance*, he contacted more than 100 publishers before he found one. Not many people would have that type of perseverance. But he did, and that's what it took.

Bell Labs

Now there's the individual approach (what works for the single person) and what works for the organization. And, of course, they're different. So let's move over to the enterprise and take a short reading. In particular, let's briefly take a look at the Bell Labs, and what they did to become a successful "idea factory" [22].

Bell Labs was unique in its conception and implementation. It was the research and development location for AT&T. At one time, there were 15,000 people employed there, to include 1,200 PhDs. That's a lot of capable people thinking about and solving difficult technology-based problems. And these were problems facing AT&T as well as the country at large. Here are just a few of the numerous areas addressed by the Labs:

1. The transistor
2. Cellular telephony
3. Satellite communications

And here are some of the notable thinkers and contributors to the scientific excellent and reputation of the Labs:

1. William Shockley
2. Claude Shannon
3. John Pierce

To recall, Shockley was the "inventor" of the transistor (Nobel Prize), Shannon created the field of "information theory," and Pierce significantly advanced the cause of satellite communications. The success as well as the sheer size of Bell Labs will likely never be seen again and fully deserves the title "idea factory." It was a laboratory that made outstanding contributions by combining superior thinking, creativity, idea generation, innovation, size, and excellent (though at times contentious) management.

Synectics

Back in 1961, William Gordon produced a book that defined "synectics" [23], a Greek word that meant "the joining together of different and apparently irrelevant elements," one that "applies to the integration of diverse individuals into a problem-stating problem-solving group." He was seriously part of a group that practiced synectics as a consulting enterprise. That enterprise was able to demonstrate that the creative process in human activity could be described and utilized in problem definition and solution. The entire theory was based also on three main hypotheses:

1. Understanding of the psychological process under which they are operating
2. Accepting that the emotional component is more important than the intellectual; the irrational more than the rational (!)
3. That the emotional irrational elements, when mastered, increase the probability of success in a problem-solving situation.

One might be surprised by the emphasis on the emotional and irrational, but Gordon makes clear that such is the case and that the continuing operation of that organization is proof that it works in a real consulting setting. In that connection, he spent some years as part of an operating group at Arthur D. Little, practicing synectics. This was, apparently, a great success, as they addressed a process involving (1) making the strange familiar and (2) making the familiar strange. In order to do that, a strong leader was needed, who was able to motivate through five key factors:

1. A shared positive experience
2. The morality of what they were doing
3. A crisis environment
4. A mutual creative responsibility
5. Pure excitement in "winning"

This entire activity appears to have been a quite successful example of finding a group process that leads, with high probability, to serious problem-solving. The possible surprise is that it has a strong dependence upon emotion and irrationality.

Barriers to thinking

As we approach the close of this chapter, we keep in mind that there are often many barriers to thinking in an appropriate and forward-looking manner. James points this out [24] by citing some of these impediments:

1. Thinking in absolutes
2. A need for immediate gratification
3. Thinking in an "either-or" fashion
4. Belief and relief in the status quo (no change necessary)
5. Just plain conformity
6. Pure gut visceral reactions, instead of due and disciplined consideration

Clearly, many of us are faced with these barriers each and every day, and try to develop workarounds so that they are as minimal as possible. This is one of the major problems of the day. For a whole host of reasons, new ideas are not accepted, and internal "voices" keep us from moving forward in whatever environment we find ourselves. Yes—it can be extremely challenging to overcome these and other barriers.

Martino barriers

Yet one more perspective on barriers to thinking has been provided by Martino [25]. He suggests that the following can often lead to poor results:

1. Misinformation accepted as true
2. Social pressure that causes negative effects and consequences
3. A vocal majority that dominates the agenda
4. A drive to agreement instead of the right answer
5. A single person that is able to subvert the overall team process
6. Special vested interests that intrude upon the group agenda
7. A premature leap to an unwarranted conclusion

This is quite a formidable list and further suggests that carrying out group processes and dynamics is a nontrivial adventure. It is not to be taken for granted.

Groupthink—revisited

There is a phenomenon that appears in a group setting, originally set forth by a Yale psychologist by the name of Irving Janis [26]. It has to do with a need for anonymity and the quiet position that members of a group will take during meetings (as well as after). It has been applied to the Space

Challenger situation as well as the Bay of Pigs misadventure. It means that one is not getting the full advantage of the thinking of every member of a group. As a result, the group process is undermined and flawed.

These same researchers have delved into the matter of Brainstorming and Brainwriting [27], both relevant to the matter of coming up with new ideas.

Nullthink

Yet another variation on the theme of thinking, or rather the failure to think clearly, is this quote from the author [28]. This pertains to another mode of thinking, defined as follows:

Nullthink—"a group process that incorporates inflexible and pebble-deep thinking, and rigid position taking, such that conclusions and decisions are impaired and very likely to be incorrect as well as unproductive."

So the reader can see—there are many ways to bring a group together and ultimately fail to puzzle their way through a problem in a satisfactory manner.

Fast and slow thinking

This area has been investigated in considerable detail by Daniel Kahneman, a Nobel laureate in economics [29], and completes this chapter. It is what one might call a completely new perspective on the matter of thinking. Kahneman sets forth a "model" for thinking that has two components— Systems 1 and 2. System 1 is fast, intuitive, immediate, and emotional. System 2 is slow, deliberate, and more logical. These two "systems" are part of all of us and they have mutual influences and uneasy interactions. Kahneman has run experiments and collected behavioral data that demonstrates the existence of these aspects of thinking. His evidence and procedures are expansive and is a real contribution to understanding this aspect of human behavior.

In an earlier work with a colleague (Amos Tversky), [29] Kahneman has proven his strength as a cognitive psychologist and an expert in the field of "decision making under uncertainty." This has shed light upon human behavior with respect to items such as human bias and risk. These have been groundbreaking activities that have indeed advanced the state of the art in this complex field.

Here are a few other areas that Kahneman and Tversky have studied and documented [29]:

Regression to the Mean—a tendency to let down after an excellent performance and make improvements after a dismal one.

Representativeness and Availability—both refer to the notion of establishing a mental model using prior experience but possibly not considering new facts and likelihoods.

Loss Avoidance—this refers to the tendency of people to keep away from sure losses even when expected value losses may be larger.

The work referred to earlier is sometimes considered to be an integral part of psychological decision theory. It is also considered to be vital to understanding how humans think and behave in a variety of situations and circumstances.

References

1. "Harvard Business Review on Breakthrough Thinking", Harvard Business School Press, *Harvard Business Review (HBR)*, 1997.
2. Amabile, T., "How to Kill Creativity", *Harvard Business Review*, 1997.
3. Leonard, D. and J. Rayport, "Spark Innovation through Empathic Design", *Harvard Business Review*, 1997.
4. Leonard, D. and S. Straus, "Putting Your Company's Whole Brain to Work", *Harvard Business Review*, 1997.
5. Morley, E. and A. Silver, "A Film Director's Approach to Managing Creativity", *Harvard Business Review*, 1997.
6. Wetlaufer, S., "What's Stifling the Creativity at CoolBurst?", *Harvard Business Review*, 1997.
7. Drucker, P., "The Discipline of Innovation", *Harvard Business Review*, 1997.
8. Lester, R., M. Piore and K. Malek, "Interpretative Management: What General Managers Can Learn from Design", *Harvard Business Review*, 1997.
9. Kim, C. W. and R. Mauborgne, "Value Innovation: The Strategic Logic of High Growth", *Harvard Business Review*, 1997.
10. Nadler, G. and S. Hibino, *"Breakthrough Thinking"*, Prima Publishing and Communication, 1990.
11. De Bono, E., *"Lateral Thinking"*, Harper & Row, 1970.
12. De Bono, E., *"Thinking Hats"*, Little Brown, 1985.
13. De Bono, E., *"Lateral Thinking for Management"*, Penguin, 1982.
14. De Bono, E., *"Super Mind Pack"*, Dorling Kindersley (DK), 1998.
15. Brown, T. and B. Katz, *"Change by Design"*, Harper Business, 2009.
16. Williams, L., *"Disrupt!"*, Pearson Education, 2011.
17. Christensen, C., *"The Inventor's Dilemma"*, Harper Business, 2000
18. Ackoff, R., *"The Art of Problem Solving"*, John Wiley, 1978.
19. Ackoff, R., *"Ackoff's Best"*, John Wiley, 1999.
20. Halpern, D., *"Thought and Knowledge: An Introduction to Critical Thinking"*, Fifth Edition, Volume 2, American Psychological Association, 2013.
21. Csikszentmihalyi, M., *"Creativity"*, Harper Collins, 1996.
22. Gertner, J., *"The Idea Factory, Bell Labs and the Great Age of American Innovation"*, Penguin, 2012.
23. Gordon, W., *"Synectics"*, Harper & Row, 1961.
24. James, J., *"Thinking in the Future Tense"*, Touchstone, 1996.
25. Martino, J., *"Technological Forecasting for Decision Making"*, American Elsevier, 1972.

26. Janis, I., *"Groupthink"*, Houghton Mifflin, 1982.
27. Sage, A. and W. Rouse, (editors) *"Handbook of Systems Engineering and Management"*, "GroupThink", p. 663, Wiley-Interscience, 1999.
28. Eisner, H., *"Topics in Systems"*, Mercury Learning and Information, 2013.
29. Kahneman, D., *"Thinking, Fast and Slow"*, Farrar, Straus and Giroux, 2011.

Exercises

4.1 What are the two points that you (like, dislike) about reductionism as an approach to problem-solving?

4.2 Name and discuss three examples of "lateral thinking."

4.3 Name and discuss three examples of "design thinking."

4.4 Name and discuss three examples of "disruptive thinking."

4.5 Cite one way to defeat each of the "Martino barriers" defined in this chapter.

chapter five

Innovation

Introduction

This chapter explores a variety of areas related to innovation, thinking, idea creation, and miscellaneous topics. Emphasis is placed upon one rather important area—innovation. We are especially interested in questions like:

- What approaches to innovation are being used and are most effective today?
- What people and institutions are committed to innovation?

Discussions of these and other related questions are the centerpiece of this chapter.

Innovation

Innovation has been defined as [1] a "process of translating an idea or invention into a good or service that creates value for which customers will pay." To this definition, we suggest that the idea or invention needs to be new, and somewhat beyond current practice.

It has been claimed that constant innovation is critical to a high-tech society and culture as we have in the United States. This is certainly true, and we see significant activities in this domain, both at the individual company level and beyond that. We proceed now by looking at one rather special company that is doing some rather special things. From there, we move on to a wider view at the national level.

A thinking enterprise

Tom Kelley documented his company's approach to innovation and conveyed many of his experiences, both within his world and with customers. This company is IDEO [2], just about synonymous with leading-edge thinking and novel solutions to a variety of client problems. Kelley's top-level approach can be quickly seen as he identifies the types of people he depends on to achieve the results he is looking for. He's given "names" of these contributors in three categories, such as (1) learning personas, (2) organizing personas, and (3) building personas.

We note that there are ten separate and distinct roles that are played at IDEO to cover the points of view that lead to innovative activities. We also recall the approach suggested by de Bono earlier with respect to "Thinking Hats." In that case, he suggested six roles. A big difference in the two approaches is that at IDEO, the people were free to shift roles when they felt it was appropriate. In the de Bono situation, the roles were fixed.

If we compare the "design thinking" concept (see chapter four) to the IDEO approach, we see two distinct and separate ways of dealing with innovation. This is to be expected and underscores the notion that there is no one way to be successful in this domain. What works in one company may not work in another. People need to find their own pathways in this endeavor.

A wider view

If we move from a single company to a look at the country at large, we can gain some insight as to how to "restore the competitive edge" [3]. In this treatise, Hage takes the approach of focusing on how to remove barriers to innovation. Briefly, the proactive version of a process of barrier removal leads to these actions:

1. Accepting and adopting new strategic opportunities
2. Broadening the views (visions) of research teams
3. Enhancing how ideas benefit from cross-fertilization
4. Integrating organizational practices
5. Improving connections in the various research arenas
6. Creating more cooperation between the public and private sectors
7. Assuring timely feedback with respect to organizational blockages

These perspectives for improvement are the author's notions for how to move forward at the national level. They make sense, and are based upon a hard look at data. How to implement these kinds of suggestions requires some level of control at the national policy level. This is often difficult to achieve.

Hage also suggests that the earlier perspectives need to be supported by a new socioeconomic paradigm. Such a paradigm would unleash "sparks of creativity and ingenuity." That paradigm also has four essential elements:

1. Predictions regarding processes of social change
2. Principles of organization that are likely to create radical innovations
3. Practices of cooperation
4. Changed perspectives with respect to economic growth

Drucker revisited

We now return to Peter Drucker who had a lot to say about innovation [4]. Basically, he believed that innovation is a discipline, and one that can be learned. Further, it is supported by a set of principles and practices. These will be cited and briefly discussed in this part of the chapter.

Drucker describes innovation as

- Systematic Innovation therefore consists in the purposeful and organized search for changes, and in the systematic analysis of the opportunities such changes might offer for economic or social innovation.
- The key word in the above is "opportunities." Drucker was clear about that point. It's all about looking for, and finding, opportunities for innovation. And once they are found, it then is a matter of analyzing them in considerable detail. Such analysis presumably verifies that the innovation brings with it positive economic and/or social benefits. A bottom line is that innovators are "opportunity-focused" and they exploit change.

To dig a bit more deeply, Drucker claims that there are seven sources of opportunities for innovation. Four of these sources are within the enterprise, with three being outside. Other points made by Drucker in his treatise on innovation include

- Purposeful innovation is based upon a search for and analysis of the opportunities
- Innovation is also based on sound concepts and precepts
- Innovation must be simple and focused
- Innovation tends to start small
- Innovation is for now, not the future
- Innovators tend not to look at risk; they look deeply at opportunity
- Innovators leave "creative destruction" in their wake
- Innovators use as their method a type of "creative imitation"
- True innovation has an economic and social effect on society

The attacker's advantage

A director from McKinsey has written a book about innovation that is most interesting [5]. One major theme is that technology is an enormous "driver" in the marketplace. Those with solid positions in a technology area must protect and support these positions, while at the same time search for new opportunities. The reality is that there are always threats to their positions, and these threats have the advantage. Why is this the case? Perhaps there are two reasons. The first is that the sources of these threats may be unknown. Secrets are hard to keep but if you want to penetrate a market held by an entrenched company, you won't be advertising that fact. Second, the timing of an attack is likely to be unknown. So the company with a lucrative business area, based upon a strong hold on some technology, must look in all directions and try to protect that line of business. Neglecting or taking for granted a "cash cow" is not a good strategy.

Another contribution made by Dr. Foster is a full explanation of the "S curve." This is a curve of performance (y axis) vs. effort (x axis) that starts out slowly (relatively flat) and then accelerates in performance with only modest increases in effort. The curve reaches a point at which better performance can be achieved only with very large increases in effort. Indeed, that leveling off may be a practical "limit" on performance. This overall curve kicks in with new technologies as they mature and penetrate the marketplace. Foster claims that you recognize it when you see it, especially when you're being attacked by a competitor. And he also claims that the attacker, most of the time, has the advantage.

Another area addressed by Foster is that of discontinuities in technology, when there is a definitive shift from one domain to another (never to return). One example is the vacuum tube to transistors transition. Another has to do with shifts from mechanical devices to computers. He recounts a set of stories that make the case—National Cash Register (NCR) in 1971, Radio Corporation of America (RCA) with electronics, and others. You will need no convincing as you recognize how devastating it can be when the technology rug has been pulled out from under you. Folks in Rochester that have depended on Eastman Kodak all these years have had to adjust to major changes in that domain.

So what do you do about all of this, if you're a company depending upon a key technology (one or more)? The difficult answer is suggested earlier. Protect what you have, as best you can, and search for new technologies and new opportunities—all the time. You need a portfolio of technologies and you need to have the people, inside your company, who will step up to the challenges that new technologies bring with them. The good news is that engineers understand technology as a rule. The bad news might well be that these same engineers might not be enthusiastic about moving from an engineering to a business orientation.

More from Harvard

The *Harvard Business Review* has a separate publication on innovation [6], an indicator of their view of its importance. Here are some ideas from that treatise.

One research area deals with how to build an innovation factory. This type of factory is a generator of ideas. A key activity is what is called "knowledge brokering" (KB). This has been systematized by following the KB cycle, which consists of these four parts:

- Obtain a set of good ideas, from a variety of sources
- Use these ideas by constant discussion and exploration
- Go after and define new ideas as derivatives and extensions of the old ideas
- Turn the best of the new ideas into new products and services

Another research area involves changes in the marketplace that lead to changes in the way that companies organize and behave. In particular, a claim is that enterprises can generally deal with evolutionary changes. They see them coming and have enough time to react. However, where they have trouble is in dealing with disruptive innovation. They don't see them coming and therefore have very short windows for reaction. This article makes suggestions as to how to cope with disruptive innovation, to include the following:

- One is looking for new processes, so create new organizational structures within the enterprise to define and implement these processes
- In addition to the earlier parts, spin off an independent entity to also create new processes
- In addition to the earlier parts, acquire a new entity to do more or less the same as the above

One might argue that the earlier remedies are overkill, but not according to the authors. They are quite serious about putting in place new capabilities to deal with significant (disruptive) change. Such changes may threaten the company or place in jeopardy an important line of business.

Finally, another research area deals with what the author calls "enlightened experimentation." Every firm tries to innovate needs to define and then run a series of experiments. The author makes the following four suggestions for how to do this:

- Reorganize a portion of the enterprise to achieve rapid experimentation
- Run a series of experiments and accept failures as they come
- Use the information gained from the earlier experiments to develop new action plans
- Develop a new experimentation *system* for rapid implementation and early forward motion

Some specific innovations

A prolific writer in and around innovation is Steven Johnson. He has sifted through this domain and offers a bird's eye view of various innovations and where they came from over time. One of his treatises suggests his "top six" innovations [7], as given later:

1. Refrigeration
2. Lenses
3. Clocks
4. Recorded sound
5. Water purification
6. Artificial light

Without taking away from Johnson's selections, this author offers his "top six" (after 1900) as

1. The automobile
2. The airplane
3. The big (atom) bomb
4. The computer (corporate and personal)
5. The transistor
6. The internet

It's curious that there is no mutual intersection between the two lists. The reader is urged to think about this matter and write down his or her "top six" list. Any new perspectives?

Yet another treatise from Johnson [8] might be considered a longitudinal investigation of new ideas and inventions since 1400. Indeed, his chronology of more than 100 key innovations starts with "double entry accounting" (1300–1400) and ends with "gamma ray bursts" (1997). Regarding new ideas, he claims that a single maxim of his book might well be that it is more important to look for connections between ideas than it is to try to protect these ideas. The former is likely to lead to new ideas, whereas the latter does not. His bibliography is useful as well as interesting if you are looking for new sources of information in regard to innovation and the creation of ideas.

The innovators

A quite prolific author by the name of Walter Isaacson has added a book to his writings, dealing with a variety of "Innovators" [9] that took part in the "digital revolution." This is a comprehensive overview of a large number of innovators, from Ada Lovelace and Grace Hopper to

Bill Gates and Steve Jobs. Here's a brief citation of a dozen special innovators and a hypothetical award the author gives them for their unique contributions:

- John von Neumann—for his brilliance in mathematics and his work on the Electronic Numerical Integrator and Computer (ENIAC) computer and computing in general
- Vannevar Bush—on his scientific writings and overall support of computing technology
- Grace Hopper—for her development of the computer compiler and support for others' contributions
- Alan Turing—for his genius and contributions to code breaking (e.g., the German Code)
- Tom Watson and IBM—for his ideas and leadership in building one of the companies that held key positions and contributed key products for many years
- Andy Grove and the Intel Team—for Grove's persistence and vision in building Intel (the company), along with Robert Noyce and Gordon Moore (of Moore's Law fame)
- Bill Gates—for his insight into the pervasive role of software and building the iconic "Microsoft" company
- Steve Jobs—for his brilliance in terms of computer design and his persistence in creating computer companies and related products
- Alan Kay and Xerox —for his leadership and inventiveness at the Xerox Palo Alto Research Center (PARC)
- Dan Bricklin—for his conception of the first spreadsheet (Visicalc), a most important success and the basis for Lotus 1–2–3 and Excel
- Tim Berners-Lee—a key player in the development of the Internet and World Wide Web
- Larry Page and Sergey Brin—for their brilliant conception of the search engine and creation of a leading edge company based upon this technology

And these folks represent just the tip of the digital revolution "iceberg." Isaacson's book is absolutely 100% engaging, just about on every page. He tells these stories with what appears to be great personal interest and enthusiasm.

Start-up nation

There's a small country in the Middle East that has a reputation for being aggressively as well as successfully innovative. That country is Israel, and it has been named a "start-up" nation by two writers that documented a number of start-ups and ongoing entrepreneurial adventures [10]. Their case for Israel as especially innovative is strong with a citation of numerous examples and anecdotes. Some of the data that they present has to do with

- Revenue levels for various Israeli companies
- Venture capital investments per capita
- Non-US companies on National Association of Securities Dealers Automatic Quotations (NASDAQ), with Israel the highest at 63
- Civilian R&D expenditure (showing Israeli leadership)

Some of the attributes of the Israelis that are suggested [10] as reasons for their success as entrepreneurs and innovators are

- Adversity breeding inventiveness
- Persistence and tenacity
- Ability to combine military and corporate experience (military and defense)
- Considerable doubt and audacity
- Constant questioning of authority
- Determined informality
- Training and intelligence
- Cross-discipline creativity
- Devotion to the overall missions
- Competitiveness

This is quite a list, especially when one considers the size of the country and the limited resources that it has at its disposal.

This author was invited to give several lectures in Israel in the year 2000. The main topic was "project management" and a few derivative subjects, all of which took about 3 days. The inviter was a school and practicing enterprise by the name of "ISEMI." This stands for—The Institute for the Study of Entrepreneurship and Management of Innovation. ISEMI was devoted to both teaching and consulting services that were connected to "entrepreneurship and innovation." Here are some ways that ISEMI represented itself:

- A formal graduate program
- Bringing the real world into the classroom
- Entrepreneurial incubator
- Seed funding and mentoring
- Helping with students' ventures and getting their degrees
- Polishing one's business plan
- Founded in 1996 by an international team of academics and entrepreneurs

I was truly impressed with their charter and the people I met in terms of their perspectives, focus, and capabilities.

Three corporate standouts

We have talked about Bell Labs [9] earlier (see chapter four) but have here a few more good words and perspectives about their dedication to innovation. We believe that innovation starts at the "top" in an organization, and Bell Labs had the benefit of special leadership in that connection. Most notable was the rise to the top by Mervin Kelly, who started out as an individual contributor and eventually became the leader that set the tone for The Labs' culture and contributions. He stressed support and connectedness between the various (and large number) smart and productive individuals. In the so-called "problem" between individual contribution and collaboration, he simply declared—we will have both. And that was a key element of success. Problem resolved by Kelly who had the clout, skills, and credibility to carry it off. And so it was that small groups of individual teams were formed and collaborated with other individual small teams to bring about breakthroughs in various fields. Example? How about the team of Shockley, Bardeen, and Brattain that invented the transistor?

Another standout R&D lab was the Xerox PARC, as of 2002 a wholly owned subsidiary of Xerox. Xerox management saw the breakthrough of copier technology and set up the Center to try for similar breakthroughs, especially in computer related fields. An inventive leader by the name of Alan Kay led the effort, resulting in many usable ideas and downstream products. Three such "products" were the graphical user interface, networking, and the "mouse." One story is that the mouse was brought into Steve Jobs' computers after he was invited to observe what was going on at Xerox PARC. Kay and his people were open in what they were doing and tried to share ideas with all that were interested. Unfortunately, there came to be a time when top management at Xerox appeared to not be listening and PARC began to fall apart. How was this manifest? Top management at Xerox began to heavily invest, for example, in real estate—a case, apparently, of not fully recognizing that they had a very good thing in the PARC. This is not a singular story. Remember the preeminence of Bell Labs, and their fall from the peaks of innovation, for a variety of reasons.

Yet another story is relevant when it comes to high performance and steadiness over a long period of time is that of the Applied Physics Lab of the Johns Hopkins University (JHU/APL). APL made significant contributions in missile defense and submarine security (one leg of our triad). Today, APL is well managed and has expanded into other national security areas. Happily, they show no signs of faltering or taking their "eye off the ball," or getting diverted in one way or another. This remains an excellent story with an enterprise that continues to do excellent work.

And continuing on with the subject of the APL, we note the commentary in the overall field of innovation represented by a comprehensive and penetrating book [11] from two employees. One important aspect of this treatise is to suggest a four-part "framework" for innovation, namely

1. Situation awareness and preparedness
2. Challenges
3. Creation processes
4. Transition outcomes

Authors of a paper within the book [12] discuss the notion of managing innovation, a key aspect of maintaining this perspective over the long haul. In this regard, they emphasize what they call the four principles of innovation: partnership, permission, privacy, and proximity. Also to be noted (from author Krill) are sets of behaviors that one would like to see from the staff and also the enterprise as well, as listed later:

For the staff
a. Awareness of important technologies, locally and across the world
b. Knowledge of capabilities that exist within the enterprise
c. Capable of meaningful collaboration with people in and out of the organization
d. Empowered to take significant actions in pursuit of innovation goals

For the organization
a. Increases in inventions and relevant publications
b. Increases in specific and important collaborations
c. Increases in potentially transformative solutions

The reader is encouraged to take a hard look at this book which is a treasure trove of information and good sense. These APL employees have done a great service to those that are thinking about and implementing innovation initiatives.

The I-Corps

A program out of the National Science Foundation (NSF) [13] has been designed to "prepare "scientists and engineers to extend their focus beyond the university laboratory, and accelerate the economic and social benefits of NSF-funded, basic research projects that are ready to move toward commercialization." I-Corps teams are formed and take part in a 7-week curriculum. This focuses on what it will take to have a commercial impact with their work. There are multiple sites that push toward transitioning their ideas and work into commercial niches and successes. There is an attempt to sustain a national ecosystem of innovation that benefits the participants as well as the country at large. Periodic Webinars provide information among and between sites and various institutions.

Most innovative companies

To keep the reader thinking about what might be going on in the marketplace, we provide here a list of the top 15 innovators of 2018 [14], as stated later:

1. Apple	6. Tesla	11. Airbnb
2. Google	7. Facebook	12. Space X
3. Microsoft	8. IBM	13. Netflix
4. Amazon	9. Uber	14. Tencent
5. Samsung	10. Alibaba	15. Hewlett-Packard (HP)

This author is most pleased to see that IBM seems to have kept pace in this important area. It's also heartening to see HP on the list, even though they're number 15.

Universities and innovation

As the last topic on the subject of innovation, we pause here to look at the ranking of universities with respect to this topic. One would expect that these institutions would maintain a sharp focus on innovation and that turns out to be the case. So here is one report that gives us the "top ten" in the world of higher education [15]:

1. Arizona State at Tempe	6. Northeastern
2. Stanford	7. University of Maryland—UMBC
3. Massachusetts Institute of Technology (MIT)	8. University of Michigan
4. Georgia State	9. Harvard
5. Carnegie Mellon	10. Duke

Perhaps the only surprise regarding the earlier list is that Arizona State ranks first, and ahead of universities such as Stanford, MIT, Carnegie Mellon, and Harvard.

As a footnote, the school at which this author taught at for 24 years (GWU – George Washington University) featured its innovative initiatives in a document called "Engineering Innovation." The innovation office has four main missions [16]: "to foster innovation, to advance education, to help students and faculty launch business initiatives, and to make connections among investors, researchers, and entrepreneurs." It is suspected that a large number of colleges and universities have similar documents and initiatives that help this country move forward in this extremely important endeavor.

A Strategy for American Innovation [17]

I.   Investing in the Building Blocks of Innovation
   -    Making World-Leading Investments in Fundamental Research
   -    Boosting Access to High-Quality STEM Education
   -    Clearing a Path for Immigrants to Help Propel the Innovation Economy
   -    Building a Leading 21st Century Physical Infrastructure
   -    Building a Next-Generation Digital Infrastructure

II.   Fueling the Engine of Private-Sector Innovation
   -    Strengthening the Research and Experimentation Tax Credit
   -    Supporting Innovative Entrepreneurs
   -    Ensuring the Right Framework Conditions for Innovation
   -    Empowering Innovators with Open Federal Data
   -    From Lab to Market: Commercializing Federally-Funded Research
   -    Supporting the Development of Regional Innovation Ecosystems
   -    Helping Innovative American Businesses Compete Abroad

III. Empowering a Nation of Innovators
   -    Harnessing the Creativity of the American People Through Incentive Prizes
   -    Tapping the Talents of Innovators through Making, Crowdsourcing, and Citizen Science

IV.   Creating Quality Jobs and Lasting Economic Growth
   -    Sharpening America's Edge in Advanced Manufacturing
   -    Investing in the Industries of the Future
   -    Building an Inclusive Innovation Economy

V.   Catalyzing Breakthroughs for National Priorities
   -    Tackling Grand Challenges
   -    Targeting Disease with Precision Medicine
   -    Accelerating the Development of New Neurotechnologies through the BRAIN Initiative
   -    Driving Breakthrough Innovations in Health Care
   -    Dramatically Reducing Fatalities with Advanced Vehicles
   -    Building Smart Cities
   -    Promoting Clean Energy Technologies and Advanced Energy Efficiency
   -    Delivering a Revolution in Educational Technology
   -    Developing Breakthrough Space Capabilities
   -    Pursuing New Frontiers in Computing
   -    Harnessing Innovation to End Extreme Global Poverty by 2030

VI.   Delivering Innovative Government with and For the People
   -    Adopting an Innovation Toolkit for Public-Sector Problem-Solving
   -    Fostering a Culture of Innovation through Innovation Labs at Federal Agencies
   -    Providing Better Government for the American People Through More Effective Digital Service Delivery
   -    Building and Using Evidence to Drive Social Innovation

U.S. innovation policy

We close this chapter with an overview of this country's innovation policy, as expressed by the National Economic Council and the Office of Science and Technology Policy in 2015 [17]. Without elaboration, the main points in this statement of policy are described in the listing that ends this chapter, as earlier.

References

1. www.businessdictionary.com/definition/innovation.
2. Kelley, T., *"The Ten Faces of Innovation"*, Currency Books, 2006.
3. Hage, G., *"Restoring the Innovative Edge"*, Stanford Business Books, Stanford University Press, 2011.
4. Drucker, P., *"Innovation and Entrepreneurship: Practice and Principles"*, Harper & Row, 1985.
5. Foster, R., *"Innovation: The Attacker's Advantage"*, Summit Books, 1986.
6. "Harvard Business Review on Innovation", Harvard Business School Press, *Harvard Business Review*, 1997.
7. Johnson, S., *"How We Got to Now"*, Riverhead Books, 2014.
8. Johnson, S., *"Where Good Ideas Come From"*, Riverhead Books, 2010.
9. Isaacson, W., *"The Innovators"*, Simon & Schuster, 2014.
10. Senor, D. and S. Singer, *"Start-Up Nation"*, Hachette Book Group, 2009.
11. Garrison, M. A. and J. Krill, (editors) *"Infusing Innovation into Organizations"*, CRC Press, 2016.
12. Paxton, L. and S. Rodriguez, "Leading Innovation into the Future", in [11] above.
13. www.nsf.gov/news/special_reports/i-corps/.
14. www.bcg.com/publications/2018/most-innovative-companies.
15. www.usnews.com/best-colleges.
16. "GW Research", *The George Washington University*, Spring 2015.
17. "A Strategy for American Innovation", *National Economic Council and Office of Science and Technology Policy*, The White House, October 2015.

Exercises

5.1 How would you give five-day instruction to a new innovation group (that reports to you)?

5.2 Write a two-page critique of Drucker's approach to innovation.

5.3 Write your own two-page explanation of "the attacker's advantage."

5.4 What are *your* "top dozen innovations" (in order of importance). Justify the selections you made and their order of presentation.

5.5 What is your own list of ten "innovators" other than those cited in this chapter? How did they innovate?

chapter six

Thinking
Miscellany

Introduction

This chapter takes a brief look at a variety of topics, all of which are related to "thinking," directly or indirectly. The topics are as follows:

- Intelligence
- Emotional intelligence
- Focus
- Artificial intelligence
- Watson the computer
- Defense Advanced Research Projects Agency (DARPA)
- *Gedanken* experiments
- Patterns of thinking
- Gelb perspectives
- Search processes

Intelligence

We presume that thinking and intelligence go hand in hand, broadly speaking. That is, the more intelligent a person is, the more he or she is able to think in a constructive way. That is probably true, but it's not the purpose of this treatise to prove or not prove this point. But it does make sense to explore the matter of intelligence in a few of its dimensions.

The conventional way of measuring intelligence is by using the Intelligence Quotient (IQ) construct, which is the mental age (using standard test questions) divided by the chronological age. There are many who do not accept this approach, for a variety of reasons. As a footnote, Einstein was reported to have an IQ of 160 (the same as Stephen Hawking), which is a healthy number. At the same time, he also failed (presumably) the entrance exam to Zurich Polytechnic. Looking behind that event, however, we discover that he was only 15 when he took that exam, and he was ahead of his class. It seems to be true that folks with high IQs have failures in their lives, for a variety of reasons. These "lapses" are part of living, but they tend to be fewer in number, the higher one's basic IQ.

It seems to be todays' common belief that the IQ measurement is flawed. The notion is that there are several forms and types of intelligence and that the IQ is too narrow. Indeed, one theory is that there are multiple intelligences. Gardener [1] claims that one can make an excellent case, for there being seven measurable intelligences, as listed later:

1. Logical–Mathematical
2. Verbal–Linguistic
3. Spatial–Mechanical
4. Musical
5. Bodily–Kinesthetic
6. Interpersonal–Social
7. Intrapersonal (Self-Knowledge)

There have also been claims, from Gardener's colleagues, that there are some 25 subintelligences. This may be true, but for our purposes here, it's time to move on in a different direction. In that case, we consider an expansion of the idea set forth by a researcher by the name of Goleman [2]. In his seminal book, he defines **emotional intelligence** as

> • abilities such as being able to motivate oneself and persist in the face of frustrations; to control impulse and delay gratification; to regulate one's moods and keep distress from swamping the ability to think; to empathize and to hope

Goleman sets this forth as a new concept, and the evidence is that many have accepted this idea and its merits. He goes on to cite Gardener's [1] work on the same basic notion, resulting in expanding these abilities into five domains:

1. Knowing one's emotions
2. Managing emotions
3. Motivating oneself
4. Recognizing emotions in others
5. Handling relationships

These are rather broad and inclusive in relation to emotions, as one can see. And the bottom line here is that not only can one have a very high IQ, but one can also have a problem with emotional intelligence such that he or she is not able to function very well. Many of us have seen this with a promising son, daughter, niece, or nephew. They're doing very well at school in terms of grades, but when they move into life experiences, things appear to fall apart. We wonder what has happened. Possibly it was due to an issue with emotional intelligence. Possibly it is there that we need to look to find reasons for "things falling apart."

After his blockbuster exposition of emotional intelligence, Goleman moved on to a discussion of "focus" [3]. In this book, he looks at what it is that promotes excellence in a variety of endeavors. He argues that we must learn to

- sharpen **focus** if we are to contend with, let alone thrive in, a complex world.

For Goleman, focus is the word that most reflects one's ability to cope and excel. He claims that there are three kinds of focus: inner, other, and outer. Inner focus deals with notions such as intuition, values, and making decisions. Other focus connects us to the people in our lives, and outer focus "lets us navigate in the larger world."

Goleman is good at digging more deeply into the components of focus as well as related concepts. For example, he realizes that "paying attention" and focus are clearly related. If you have trouble paying attention, you'll also likely to have trouble focusing. Then he parses the notion of attention and gives us a richer, more penetrating analysis.

Goleman explores the classic model of the stages of creativity. His conclusion? There are three modes of focus:

- Orienting
- Selective attention
- Open awareness

Goleman also spends sometime looking at the phenomenon of groupthink. You will recall that this has to do with how people interact in a group situation, often coming to quite poor decisions. Some of the features of groupthink, he concludes, are

- A shared self-deception of the participants
- The unstated assumption that the group knows everything it needs to know
- Unwillingness to speak up and endure the risks that such might bring about

A kind of bottom line for this treatise is simply that people who are able to focus and also able to embrace what might be called "smart practice": (1) can improve their habits, (2) add new skills, and (3) sustain excellent performance and interactions with others. All of these require embracing change, which we know is difficult to do.

Artificial intelligence

In 1988, some 30 years ago, this author published a book [4] with a separate chapter on artificial intelligence (AI). The main ideas featured in this chapter were as follows:

- Knowledge-based expert systems
- Using the aforementioned for advanced diagnostics
- New languages and how they facilitate AI (e.g., LISP—LISt Processing language)
- Logic and truth tables
- Inference
- Software for AI
- Available (commercial) expert systems

Moving from 1988, we see three noteworthy points. The first is that interest in expert systems appears to be still very strong. More about that is given later. Second, there has been a massive increase in computing power and storage in these 30 years (recall Moore's Law). This is no surprise. And finally, we see real systems being built that could fall within the purview of AI, although they tend not to be. Two examples are (1) the most advanced search engines (as per Google) and (2) the Watson-type computers that handle English language inputs. For both of these examples, the technology is extremely fast and also accurate.

On the matter of expert systems, we take note of the fact that people are still working on them. They have improved the two major components, namely, the knowledge base and the inference engine. If we add to the mix a major improvement in computing power and memory, we have a rather powerful piece of technology that can readily be applied to problem-solving today. A field highlighted by this author is certainly medical diagnostics. This can be a specific way to supplement what an individual doctor has to offer. And the expert system is able to add additional information, such as the likelihood that any particular diagnosis applies to the case in question. Other areas of application include investment tactics and strategy, petroleum engineering, and cancer research.

As far as the relationship between AI and what systems engineers do, from day to day, there are two areas to mention. One is the fact that systems engineers are being called upon to build such systems, and they need to be up to speed on the latest in AI systems and technology. The second is that the AI technology may be part of systems that are being considered, such as inference engines and search algorithms. Perhaps, the reader can think about other connections between AI and systems engineering.

Watson (the computer)

Watson is an IBM computer named after Thomas Watson, Sr., IBM's first Chief Executive Officer (CEO), and father to the man that had enormous positive influences on that iconic company. It came into prominence when it was competing on *Jeopardy!* against the two strong winners, namely, Ken Jennings and Brad Rutter. Yes, indeed, the winner of the competition was the machine, the Watson (super) computer.

Watson operates by considering the *Jeopardy!* format—questions set forth in natural language. These are the "inputs," and Watson had access to 200 million pages of structured and unstructured content. The question was—could Watson search, retrieve, and make inferences so as to answer the questions better than Jennings and Rutter, meaning with greater accuracy and response time (speed)?

So, this spectacular performance would lead one to consider what would be considered the ultimate question:

- Is Watson "thinking"?

  The basic answer is that Watson is carrying out several of the functions of the brain, but is not really thinking. It's doing something that mimics thinking in some ways, but is not able to do everything our brains are capable of doing.

  Watson continues to be a strong asset of IBM, and the company makes it available to selected clients to tackle difficult and important problems. Students of AI and advanced thinking machines should take a hard look at the architecture of Watson, and what might be done to continue to make advances in this field.

  Many might disagree with this author and instead claim that Watson was exercising at least some form of "thinking." Among that group might well be several researchers at a research consortium within the MIT Media Lab, where they have been looking at what they call "TTT" (things that think) [5]. Here are some ideas in the source that give us some insight into areas they appear to be investigating:
  - Books that change into other books
  - Special musical instruments that help the learner do better
  - Shoes that communicate by means of body networks
  - Printers that provide working (vs. static) things
  - Money that contains behavior as well as value

A brief note about DARPA

DARPA has been known to this author for many years. Recalling a face-to-face conversation with several DARPA researchers, I noted that they made a point out of being not only advanced, but also beyond what one might call leading-edge research. To stretch the metaphor, they were "beyond the beyond." And at the time, I considered that position quite appropriate—someone had to be in that place, and why not DARPA?

So with respect to intelligence, and the broad field that it defines, DARPA is "building a brain" [6]. They claim that, in the last decade or so, "computing and data are becoming the central tools for innovation." This is not surprising. The power of computing has increased dramatically, and the data that has become available has done the same. A big question has been—are we able to handle the "big data" and make something special out of it? The answer to that question has become an emphatic "yes."

An example of where DARPA seems to be here, in the 2018 time frame, is its SIMPLEX (Simplifying Complexity in Scientific Discovery) project. The purpose of this project is to "create new mathematical and algorithmic abstractions for data-driven discoveries." Application areas for this work have included

- Genomics
- Anthropology
- Materials science

In many of DARPA's programs, they are "creating technologies to enable new forms of human–machine problem solving." This is a very exciting research, promising to offer breakthroughs in the connection between man and machine, and how they may be able to work together in ways never seen before. Keep in mind that the perspective of DARPA is to be thinking not about what might be just around the corner in terms of technology advances. They are ways beyond that corner, generating ideas about the distant future.

Gedanken *experiments*

The word *"gendanken"* means "thinking," and this approach apparently was coined by Einstein as he considered new concepts in physics and the creation of his perspectives on relativity [7]. One such experiment involved thinking about riding a light wave looking at a light wave moving in parallel. The constancy of the speed of light gave him insight that applied to his special relativity theory. Another *gedanken* experiment can be stated, using his words:

> - I was sitting on a chair in my patent office in Bern. Suddenly a thought struck me. If a man falls freely, he would not feel his weight. I was taken aback. This simple thought experiment made a deep impression on me. This led me to the theory of gravity.

So it was no less than Einstein that showed how powerful "thinking" can be. Of course, this was Einstein, one of our most notable thinkers. But perhaps, we can take a lesson from him and run our own *gedanken* experiments. Take another look at one of the problems you've been considering. Is there a *gedanken* experiment that you can run? Is there a scenario that sheds light on the problem? Remember, this is all in your mind—you don't even have to get up off the couch.

Patterns of thinking

A project carried out at the Harvard Graduate School of Education [8] shed some light on the "understanding, teaching, and assessment of thinking dispositions." In this investigation, emphasis was placed upon the latter term—thinking dispositions. They had identified people that are "disposed" to taking the correct path in terms of patterns and sequences of thinking. This type of behavior is supported by three distinct components:

- Ability
- Inclination
- Sensitivity

The latter two have made unique contributions to intellectual behavior. Sensitivity has been a "chief bottleneck" in terms of having the right disposition to solve problems and think through a complex situation.

A Gelb perspective

Michael Gelb has contributed significantly to understanding how Leonardo da Vinci thought about the world around him. His penetrating book [9] covers a lot of ground and reinforces the idea that da Vinci was indeed one of our greatest thinkers. Gelb winds up focusing on seven principles of da Vinci:

1. Curiosity and quest for learning
2. Testing knowledge through experience and learning from mistakes
3. Continual refinement of the senses
4. Embracing ambiguity, paradox, and uncertainty
5. A balance between science, art, logic, and imagination
6. Cultivation of grace, fitness, and poise
7. Systems thinking

We note the broad range of thinking suggestions, revealing what might be called a "whole-brain" approach. Da Vinci was able to master all of the earlier principles and fully integrate them as he proceeded with his life and his substantial contributions. We also note the last item in the earlier list—systems thinking. We've seen that notion before (in this treatise), and see that it's not new. It goes back to da Vinci's time—the 1400s and 1500s.

Finally, Gelb articulates what he considers to be the key attribute of today's renaissance man or woman:

- Computer literate
- Mentally literate
- Globally aware

These three are critical, along with a strong knowledge of classical liberal arts.

Search processes

A large number of problem areas that require our best efforts at thinking involve some type of search process. Here are some examples, as we are looking for:

- A better approach to a cure for cancer
- A missing boater in a search and rescue operation
- Incoming missiles that represents threats
- The best architecture for a large-scale system of systems
- That proverbial "needle in a haystack"

We have the classic questions to deal with—where do we search, how do we search, how much effort should go into each level of search, what are the likely search features, and so forth.

So we end this chapter with a search question for the reader. This will require some "heavy-duty" thinking, as the solution is not trivial. The full answer can be found in one of the author's books [4, page 183]. Here is the (coin-weighing) problem:

- You have 12 coins, all of which are the same, but one is either heavier or lighter than the others. You also have an equal arm balance, and nothing more. In exactly three weighings, determine the coin that is different and whether it is heavier or lighter than the others.

References

1. Gardener, H., *"Frames of Mind"*, Basic Books, 1983.
2. Goleman, D., *"Emotional Intelligence"*, Bantam Books, 1995.
3. Goleman, D., *"Focus: The Hidden Driver of Excellence"*, HarperCollins, 2013.
4. Eisner, H., *"Computer-Aided Systems Engineering"*, Prentice-Hall, 1988.
5. Gershenfeld, N., *"When Things Start to Think"*, Henry Holt, 1999.
6. Regli, W., *"Building DARPA's Brain"*, *Expert Opinion, IEEE Computer Society*, January 2018.
7. Perkowitz, S., "Gendankenexperiment", *Science*, www.britannica.com/science/Gedankenexperiment.
8. "Patterns of Thinking", Harvard Graduate School of Business, *Project Zero*, http://gse.harvard.edu.
9. Gelb, M., *"How to Think Like Leonardo da Vinci"*, Dell Publishing, 1998.

Exercises

6.1 Write two pages that elaborate upon Goleman's approach to "focus."

6.2 Suggest your own *"gedanken"* experiment that is relevant to the topics in this chapter.

6.3 Write two pages in which you compare da Vinci's approach to "systems thinking" and an approach that you select from another person cited in this book.

6.4 Write a page about one or more projects at DARPA that is relevant to the themes in this chapter.

6.5 Solve the problem at the end of this chapter (coin weighing).

chapter seven

Engineers with special thoughts

Introduction

In this chapter, we take a brief look at contributions made by modern-day engineers, scientists, and mathematicians. They were able to "think outside the box" and bring notable systems and companies into being through their insights, specific ideas, and perseverance. Some became businessmen and were able to promote technology and innovation for most of their lives with sizable monetary contributions. Others have gotten on the lecture circuit and shared their wisdom and advice with the rest of the country and world. All of them had special thoughts and special stories to tell.

Chester Carlson

This is a person perhaps not well known to lots of folks. Basically, he was a researcher that held key patents in the field known as electrophotography. He also worked under the umbrella of Battelle Memorial Labs, in the 1940s. Joe Wilson, the president of an enterprise by the name of the Haloid Company, was a supplier of photographic paper, largely to Eastman Kodak. Both had their headquarters in Rochester, New York. Wilson was looking for ways to expand his business and found out about Chester Carlson and his work as well as his patents. Wilson also became convinced that with the Carlson patents, as well as Carlson himself, he could take the next step and bring copiers into the mass market. This was, as they say, a risky notion, sometimes described as "betting the company." But that's exactly what Wilson did when he got together with Carlson (and many others) to bring copier technology to the "masses." Behind that considerable step forward was Carlson the man and Carlson, the key employee of what ultimately became the Xerox Company.

On a more personal side, Carlson was born into poverty in 1906 and came to the United States from Sweden. With great perseverance and intelligence, he was accepted at the California Institute of Technology (Cal Tech) as a junior. From 1935 to 1937, he searched for a way to produce a copier, trying to combine electrostatics and photoconductivity. He received the first of four patents in 1942 and made a deal with Battelle Labs. They thought it was a magnificent invention but that it could not become a product. Joe Wilson stepped in with his vision and determination as well as his leadership abilities and made it all happen. The reader is referred to the account by Ellis [1] for a more complete story about Carlson and Xerox as well as other key players in the company's evolution.

Norman Augustine

Norman Augustine started out as an aeronautical engineer and wound up running one of the largest companies in the country—Lockheed Martin. He has also been a trusted advisor on important national issues that have to do with technology and management, and a college professor par excellence with at least two books to his credit. Dr. Augustine, definitively both right- and left-brained, is one of those natural leaders who is knowledgeable and balanced, and who can move from one field to another with apparent ease as well as success.

I have especially enjoyed two of his books [2,3]—one dealing with his "laws" and the other with his "travels."

Moving to Augustine's travels [3], we find him ranging far and wide on a variety of subjects, each being of great interest to the scientist, manager, and executive. The flyleaf of this book suggests emphasis on the following: ethics, leadership, competiveness, mergers and acquisitions, and reengineering. He expands on the matter of leadership, putting forth five attributes of a leader, namely

1. Inspiration
2. Perseverance
3. Courage
4. Selflessness
5. Integrity

In his usual easy style, he has much to say about each of these. And on his own overall theory of management, he says it can all be summed up in 14 words—"Find good people, tell them what you want, and get out of their way." That's probably the most succinct and useful advice anyone will offer on this otherwise verbose arena, with the possible exception of another suggestion that he makes. And that is—make sure to learn from your mistakes. They are inevitable and we all make them. The trick is to come away from each of them with a distinct lesson learned.

As one might expect, there are numerous high-level achievements in his career. Here are some of them:

- CEO of Lockheed Martin and of Martin Marietta
- Undersecretary of the Army
- Chairman of the National Academy of Engineering
- Trustee of John Hopkins and Princeton Universities

- Professor at Princeton University
- Awarded 13 honorary doctorate degrees
- National consultant to the Department of Defense (DoD), National Aeronautics and Space Administration (NASA), and other agencies

Norman Augustine has had a most notable and varied career, and it appears that he has not finished with his service and contributions to this country.

Jack Welch

Dr. Welch came up through the ranks and eventually became the president of General Electric. Originally a chemical engineer, he steered that company through good and bad times with fierce determination and skill. Along the way, he earned a reputation as "neutron jack," as he "blew up" parts of the company that were no longer productive, by his "objective" measures [4]. By blow up we mean that, for each significant business area, his overall strategy was to either "invest, repair, or sell." He was constantly searching to be, or become, number one or two in any given line of business (LOB). Mostly, that covered the waterfront. Here are some of the facts and initiatives embraced and reinforced:

- Installed as chairman and CEO of GE in 1981
- Retired after 20 years
- Built more than $450 billion in capital value during his tenure
- Instituted a "fix, sell, or close" perspective across the company, with great consistency
- Emphasized globalization, six sigma quality control and assurance, movement to greater degrees of services, and greater e-business initiatives

In another book (with his wife Suzy), Welch confronts "74 of the toughest questions in business today" [5]. Many of these require new patterns of thinking, and acceptance of the basic notions of change. Here are some of the key issues discussed in this treatise:

- Major support for "continuous improvement" vs. new large-scale initiatives
- Encouraging five traits for our new leaders, namely: positive energy, energizing others, a competitive and competent edge, ability to execute and a passion for the task
- Being not tough, but tough-minded
- Fixing organizational lacks, for example: lack of an inspirational mission, lack of a clear set of values, and lack of a rigorous appraisal system
- Building trust

- Motivating
- Encouraging internal entrepreneurial behavior

A direct link can be established between the last of the earlier items and innovation within the enterprise. That's a main linkage between this Welch book and many of the themes in this book dealing with thinking and generating new ideas. Welch has not lost his touch and his understanding of what might be called the fine art of management.

Eberhardt Rechtin

Eberhardt Rechtin was what this author would have to call a Master Engineer, with excellent technical and management skills. He too was both right- and left-brained, and published at least three books [6–8] that are on my bookshelf and that I go back to read and reread.

Dr. Rechtin was trained as an electrical engineer and made his mark early in terms of the design and implementation of the Deep Space Network. This was a large and important project out of the Jet Propulsion Lab in Pasadena. Dr. Rechtin moved on to become director of DoD's Defense Advanced Research Projects Agency (DARPA) and also an assistant secretary of defense for telecommunications (1972–1973). Later, he was a chief engineer for Hewlett-Packard and president of the Aerospace Corporation. The latter entity was an important advisor (think tank), principally for the Air Force. He also served as a professor of engineering at the University of Southern California (USC).

His book on system architecting [6] remains a seminal treatise on the subject. He used his architecting principles and heuristics to great advantage in his teaching, lecturing, and consulting as well as his management positions. These were all significant contributions to the field.

He received many awards for his service and received his doctorate in electrical engineering from Cal Tech. He passed away at the age of 80 in 2006.

Andy Grove

Andy Grove was born Andras Grof in 1936 in Budapest, Hungary. He came to the United States in 1956, went to the City College of New York, and graduated with a degree in chemical engineering. He continued on with his studies and earned a doctorate in 1963. He was a technical person who spent a lot of time figuring out how to manage a technical enterprise and became very good at it [9]. His management style was described as "constructive confrontational." That meant that he would relentlessly attack problems but with a soft touch. This led to enormous successes, and a strong following from all of those around him. He always looked at management as a team activity and was excellent at building teams.

He supported the ideas of creating leverage in the organization and making the most of meetings of various types. Each meeting had a purpose, and high output objectives were sought and achieved. He set forth an approach to decision making that was basically composed of three steps:

- Free discussion
- A clear decision
- Full support

Several iterations of these steps were called for, when necessary. In terms of planning, he was clear about the steps of (1) environmental demand, (2) determining present status, and (3) what to do to close the gap. He also used the Management by Objectives (MBO) process, and his advice about personal assessments consisted of three steps: Level, Listen, and Leave Yourself Out (his three "L's). He used and refined the concept of "task-relevant maturity". He became well known for his perspective about management embodied in the phrase "only the paranoid survive." He is also well known for his formula:

- A Manager's Output is equal to the Output of His Organization + the Outputs of Other Organizations Under His Influence

He had the respect of many high-level managers in other organizations, including competitors. That's reflected, in part, by his selection on the cover of *Time* magazine as "Person of the Year." John Doerr, a serious technology pioneer, declared that "Andy Grove was a towering leader, mentor, and educator—he was ruthlessly, intellectually honest, and rightly proud of building Intel" [10].

He took a job at Fairchild Semiconductor, where he apparently learned the semiconductor business, top to bottom. That left him in an excellent position to move on to Intel, hired by Robert Noyce and Gordon Moore. He became president of Intel and served from 1979 to 1997. He then became chairman of the board from 1997 to 2005. During his tenure as president, Intel revenues soared from $4 billion to $197 billion. This represented the best performance in the complex semiconductor arena, where he developed the popular 386 and Pentium chips.

He received many awards and gave back to his college by means of a $26 million grant to City College of New York (CCNY), renaming the School of Engineering as the Grove School of Engineering.

Irwin Jacobs and Andy Viterbi

Here are two names possibly not too well known in the engineering community. But they both had wonderful careers and made substantial contributions as engineers and innovators.

As a curiosity of history, both Jacobs and Viterbi were in the same place at the same time, but apparently did not cultivate a connection at that time. This was in 1956–1957 in the electrical engineering department of Massachusetts Institute of Technology (MIT). Both were graduate students at the time. From there, they moved in different directions, only to connect later in their lives.

Irwin Jacobs had a major influence on a particular niche in the field of communications. Here's an overview of his background and accomplishments. He completed his doctor of science studies in 1959 and became a professor at MIT. An early accomplishment was the writing and publishing of a seminal text in communications engineering. His strong academic career took an interesting turn when he met Viterbi in 1963 at a conference. He went off to work at Jet Propulsion Laboratory (JPL) for a year (1964–1965) and soon connected with Viterbi in the founding of a company they called Linkabit. This was a great success, moving from Louisiana to San Diego and spawning many companies in the telecommunications field over a period of some 25 years. One of these was Qualcomm [11].

Meanwhile, let's follow the path of Andy Viterbi as he moved to Louisiana and joined the staff of JPL in Pasadena as a senior communications engineer. He worked intensely on unmanned satellites, and in parallel with his work, obtained a doctorate in engineering at the USC in 1962. At that point, he constructed what came to be known as the "Viterbi Algorithm." This was a breakthrough and dealt with digital communications decoding. It served an important function in terms of eliminating signal interference. Dr. Viterbi was also a codeveloper of a communications technique known as Code Division Multiple Access, which is used in the cell phone world. For his many contributions, Andy Viterbi has received numerous awards from a variety of technical organizations.

Returning to the matter of Qualcomm, this was a company started by Jacobs and Viterbi in 1985 (and a few others). It was very successful in the fertile fields of satellite communications and digital wireless devices. In 2017, a company by the name of Broadcom came along and tried to take over Qualcomm, offering $103 billion. This was rejected by Qualcomm's board, a decision supported by Irwin Jacobs. It's not clear as to the status of this possible transaction as of the publishing of this book.

Simon Ramo [12]

Simon Ramo was literally a giant in both technology as well as a company that builds and provides that technology. He is known to be the major driver in this country's Intercontinental Ballistic Missile program, installing its center of gravity in Los Angeles and environs. He proceeded with high-tech weapons, especially as to his involvement and role with TRW (the "R" being Ramo). He also worked with great success at Hughes, GE, and Bunker-Ramo.

At 23, he received his PhD with majors in both physics and electrical engineering. He retired in 1965, but continued to contribute to many study groups and national investigations. He also served as a member of the faculty of the USC, a powerhouse of technical talent and expertise. He received many awards including the Medal of Freedom from President Reagan. I counted 20 others medal and awards in an article citing his accomplishments.

He was a strong influence on defining and carrying out "systems engineering," both through his writings and also the TRW company. The company was a preeminent systems engineering enterprise with a reputation that preceded it.

He also thought about and commented on the "systems approach" in a classic joint paper [13] with a Robin St Clair. When he had something to say that he thought was important, he wrote a book about it. Here are a few (from many more) illustrations of books to his credit:

- *The Management of Innovative Technological Corporations* (1980)
- *What's Wrong with Our Technological Society and How to Fix It* (1983)
- *Tennis* by Machiavelli (1984)
- *The Business of Science: Winning and Losing in the High-Tech Age* (1988)
- *Tales from the Top: How CEOs Act and React* (2011)

He passed in June of 2016 at the age of 103.

Claude Shannon

Claude Shannon spent a considerable number of years with the iconic super think tank known as Bell Labs. So we have a reasonable account of some of his activities in an in-depth look at Bell Labs from Jon Gertner [14]. That turns out to be the case, and some of the adjectives used in that book to describe Dr. Shannon were as follows: retiring, eccentric, diffident, amiable, whip-smart, whimsical, and prankish.

Indeed, Claude Shannon was brilliant but somewhat eccentric, as he apparently moved through the halls of Bell Labs on a unicycle, carrying out the difficult task of juggling. And unlike many of his colleagues, he disappeared into his office, closed the door, and spent hour upon hour doing—whatever. The presumption was that he was thinking and working on significant problems on behalf of the lab. And whatever the optics, he came up with an approach to communication theory that was a real breakthrough. That took the form of creating the basis for "information theory." Indeed, he became known as the "Father" of that field. Here are two quotes from Gertner's book that reinforce that point:

- Bell Lab colleagues described it as one of the great intellectual achievements of the 20th century.
- I know of no greater work of genius in the annals of technological thought (from Bob Lucky of Bell Labs).

Shannon had worked out a new theory and published the results in 1948. He was also creating the new field of information theory, the basis for extensive work on coding for digital communications for years to come. At its core, it was built upon the notion of entropy (H), defined as

$$H = -\sum p(i)\log p(i)$$

where the p(i) are the probabilities of the occurrence of transmitted symbols or letters.

So, the whole idea of modern communications was viewed in probabilistic terms, and the resolution of uncertainty regarding what symbols were sent and received. This breakthrough idea was profound, and deserved the accolades and recognition it received.

Shannon's paper at Bell Labs was reproduced in book form [15], which covered the key subjects, several of which are listed later:

- Discrete Noiseless Systems
- The Discrete Channel with Noise
- Continuous Information
- The Capacity of a Continuous Channel
- The Rate for a Continuous Source

We are truly indebted to Claude Shannon for his penetrating and brilliant contributions to communication theory and systems and for creating the field of information theory.

Jay Forrester

Jay Forrester formulated a new and powerful modeling theory and technique, otherwise known as system dynamics [16]. Derivatives of this basic work were industrial dynamics, urban dynamics, and world dynamics. It has been instantiated in a commercial package by the name of DYNAMO.

Forrester's development of system dynamics apparently served as a foundation for the Club of Rome work, and the results were provided by Dennis and Donella Meadows [17] in that connection. Thus, we see an immediate and important application of Forrester's contribution. Many others are reported in the literature, in many domains to include economic and social science. His broader interest in "systems" was suggested by his book *Principles of Systems* published in 1968.

Forrester's strength and capability in terms of building the system dynamics model was likely a derivative of his work on aircraft flight simulators and air defense systems (such as SAGE—Semi-Automatic Ground Environment). He became an expert in these types of systems, and recognized that high-fidelity simulations were here to stay and that he could make an important advance in this arena.

Jay Forrester was trained as an electrical engineer and received his BS degree in 1939. From there, he went off to MIT and basically stayed there for his entire career. He moved within MIT from electrical and computer engineering to the Sloan School of Management. In a speech at the System Dynamics Society in Germany [18], he "revealed" that "another reason for moving to management was that I was already in management." Clearly, he was comfortable with both highly technical as well as management problems and issues. Here we have another person who used both his right and left brains with great ease and facility.

Forrester was widely recognized for his outstanding contributions. Examples are two awards that he received, namely, the IEEE Medal of Honor (1972) and the IEEE Computer Pioneer Award.

Tom Watson, Jr.

Often, Tom Watson, Jr., was the elephant in the room. He was the major driver in the building of IBM over the years, and IBM was a leader over these years despite its several ups and downs. As such, Watson had enormous power and influence in the industry, respected for his management acumen and his understanding of how to build a company.

Tom Watson, Jr., had a father who was also deeply connected to IBM. In effect, the elder set the stage for his son who made the most of this connection and opportunity. His tenure as CEO at IBM was during the years 1952–1971, a period of time when he brought IBM into a most prominent position with large mainframe computers. Familiar names include the IBM 7070, 7090, 1620, 1401, and the System/360 series. The latter was considered, at that time, a serious gamble in terms of its cost and potential place in the market. It turned out to be a great success. This type of success led *Fortune* magazine to declare that Watson was "the greatest capitalist in history" [19]. He was also awarded the Presidential Medal of Freedom from Lyndon Johnson in 1964 and other medals of recognition.

He was a master agent of change and of formulating effective organizational units and structures. He received credit for bringing IBM into its preeminent position on top of the computer industry. He led the charge in revenue increases, apparently moving into the multibillion dollar domain with relative ease, fairly early (in 1958). Two of his passions were sailing and piloting. He donated millions to support causes he found worthy, including tens of millions to Columbia University. He passed away in December of 1993 at the age of 79, and clearly had accomplished a lot during his years.

Frederick Brooks, Jr.

Frederick Brooks is a master software engineer with both practical and theoretical background. Since software has become such a critical part of most large-scale systems, we reserve this space here to recognize Dr. Brooks, and another outstanding software engineer at the end of this chapter.

Frederick Brooks is perhaps best known for his role in developing both the hardware and the software for the IBM System/360 series of computers. He led the charge in what might be described as the most critical challenge in the computer industry at that time. There was a lot of risk associated with that project, and Dr. Brooks overcame the risk and brought great success and financial returns to IBM during an important time in their history.

His book, *The Mythical Man-Month*, has become a classic in the field [20]. Among other topics, Brooks articulates several "laws" and perspectives that work in the domain of software engineering. Here are some examples:

* The man-month, as a measure of the size of a job, is dangerous and deceptive
* Adding more people to a job lengthens, rather than shortens, the schedule
* Many key projects fail because of failures in communication and organization (rather than difficulty with the code)
* Representation is the essence of programming
* Software systems should be built by incremental development

The last-cited bullet earlier implies that one needs to embrace evolutionary vs. revolutionary development. He is also clear about supporting Niklaus's top-down design precepts and procedures and provides advice on all significant aspects of building software systems.

In a second book [21], Brooks emphasizes the design process itself, strongly arguing that it too must be "designed." Of course, this is an important point—software design is clearly not a random process. Here are a few of the other topics he deals with in this book:

* Rationalism vs. empiricism in design
* Collaboration in design and two person teams

- Design as a search process
- The persisting rational model
- The worst computer language ever

Dr. Brooks set up the computer science department at Chapel Hill, North Carolina, and served as it chair from 1964 to 1984. He has received the National Medal of Technology and the Turing Award from the Association for Computing Machinery. His consultancies have included the Defense Science Board and the National Science Board. He has clearly influenced, in a most positive way, at least two generations of software engineers in both computer architecture and software.

Barry Boehm

Barry Boehm, as of this writing, is an active and significant contributor to the fine art of software engineering. Dr. Boehm was trained as a mathematician (PhD from University of California in Los Angeles (UCLA) in 1964), and is well known for his COnstructive COst MOdel (COCOMO) I and II initiatives. These were embodied in two seminal books [22,23]. Both books represent major advances on how to estimate the costs of software projects. They are crucial sources for the software engineering community, especially for large-scale software projects.

In addition to COCOMO, he is credited with having first defined the software "Spiral model" and the "Theory W" (win- win) approach to building and managing software programs and projects.

Dr. Boehm serves as the TRW professor of software engineering at the USC Center for Systems and Software Engineering. He is also the chief scientist in a DoD Engineering Research Center (working with the Stevens Institute of Technology). He also held positions at DARPA and Department of Defense – Research and Engineering (DDR&E), both parts of the Department of Defense. He has received numerous awards for his many groundbreaking contributions. He is truly one of our national treasures in software systems and engineering.

References

1. Ellis, C., *"Joe Wilson and the Creation of Xerox"*, John Wiley, 2006.
2. Augustine, N., *"Augustine's Laws"*, American Institute of Aeronautics and Astronautics (AIAA), 1982.
3. Augustine, N., *"Augustine's Travels"*, AMACOM, 1998.
4. Welch, J., *"Jack: Straight From the Gut"*, Warner Books, 2001.
5. Welch, J. and S. Welch, *"Winning: The Answers"*, Collins, 2006.
6. Rechtin, E., *"Systems Architecting"*, Prentice-Hall, 1991.
7. Rechtin, E. and M. Maier, *"The Art of Systems Architecting"*, CRC Press, 1997.

8. Rechtin, E., *"Systems Architecting of Organizations: Why Eagles Can't Swim"*, CRC Press, 2000.
9. Grove, A., *"High Output Management"*, Vintage Books, 1983.
10. See Andy Grove's obituary with a Google or other search engine.
11. West, J., "Before Qualcomm", www.sandiegohistory.org, December 29, 2008.
12. Ramo, S., *"Conquering Complexity"*, The Defence Engineering Group, University College London, #10, 2005.
13. Ramo, S. and R. K. S. Clair, www.incose.org/productspubs/doc/systems approach.pdf.
14. Gertner, J., *"The Idea Factory: Bell Labs and the Great Age of American Innovation"*, Penguin Books, 2012.
15. Shannon, C. and W. Weaver, *"The Mathematical Theory of Communication"*, The University of Illinois Press, 1949.
16. www.technologyreview.com/s/538561.
17. Meadows, D., *"Thinking in Systems"*, Sustainability Institute, Chelsea Green Publishing, 2008.
18. Forrester, J., "The Beginning of System Dynamics", *Banquet Talk, System Dynamics Society, Stuttgart, Germany*, July 13, 1989.
19. "The Greatest Capitalist in History", *Fortune Magazine*, August 1987.
20. Brooks, F., *"The Mythical Man-Month: Essays on Software Engineering"*, Addison-Wesley, 1975/1995.
21. Brooks, F., *"The Design of Design: Essays from a Computer Scientist"*, Pearson Education – Addison Wesley, 2010.
22. Boehm, B., *"Software Engineering Economics"*, Prentice-Hall, 1981.
23. Boehm, B., et al., *"Software Cost Estimation with COCOMO II"*, Prentice Hall PTR, 2000.

Exercises

7.1 Write your own two-page "story" about an engineer that deserves to have his or her story told.

7.2 Which of the "stories" in this chapter do you find most compelling? Why?

7.3 Write your own two-page "story" about a software engineer who deserves to have his or her story told, other than the two software engineers in this chapter.

7.4 Write your own two-page "story" about a systems engineer who deserves to have his or her story told, other than the people cited in this chapter.

7.5 Who can you cite that made the most significant contribution to the field of "thinking." Explain that contribution and why you find it significant.

chapter eight

Top dozen suggestions for systems engineers

One of the main purposes of this book is to place various perspectives regarding the fine art of thinking closer to the domain of systems engineers. There is an expansive literature on systems engineering (much provided by International Council on Systems Engineering (INCOSE)). There is also a huge amount of information available on "thinking" and related topics. Again, this book is an attempt to make "thinking" information more readily accessible to the systems engineer. This author hopes that, by so doing, the systems engineer will have a broader and more useful set of ideas as to how to perform the complex and multifaceted tasks of systems engineering. In this chapter, we look specifically at the "top dozen" suggestions for what the systems engineer should pay special attention to, as recommended by this author who has spent considerable time as part of the systems engineering community.

Embrace the "systems approach"

The "systems approach," as defined by the author, is articulated in chapter three. It is embodied in the following 12 suggestions:

- Establish and follow a systematic and repeatable process
- Assure interoperability and harmonious system operation
- Consider alternatives at the various steps of design
- Use iterations to refine and converge
- Create a robust and slow-die system
- Satisfy all agreed upon user/customer requirements
- Provide a cost-effective solution
- Assure system sustainability
- Use advanced technology, at appropriate levels of risk
- Consider all stakeholders and their concerns about the system
- Design and architect for system integration
- Employ systems thinking

Each of these items, in terms of potential implementation, requires a serious commitment. This is not easy to do but will provide serious benefits.

Practice "systems thinking" (see the aforementioned last item)

"Systems Thinking" has been addressed by Peter Senge [1] and few other investigators. Senge introduced systems thinking as "the Fifth Discipline" needed to build and sustain a "learning organization." The five disciplines are as follows:

1. Personal mastery
2. Mental models
3. Team learning
4. Building shared vision
5. Systems thinking

Senge also defines systems thinking as the discipline that integrates the disciplines, fusing them into a coherent body of theory and practice. In chapter three, we have described systems thinking in terms of factors such as

- Holistic
- Integrated
- Generalized
- System-wide
- Fusion
- Top-level

Consider hybrid thinking approaches

A "hybrid" approach is some combination of "thinking up," "thinking down," and "thinking laterally." Thinking up is what we do when we consider the problem in a broader context, as part of a larger "system." Thinking down is what we do when we "drill down" within a limited domain to get more data and press forward in a single solution space. Thinking laterally is what we do when we move sideways to explore the features of an attractive system. You will recall that de Bono was the creator of lateral thinking [2]. So, it may turn out that, for any new system development, the design team wishes to employ a combination of all three approaches. That is perfectly acceptable as long as there is a rationale for doing so.

Question conventional wisdom

There are many parts of systems engineering and problem-solving in that domain that need to be challenged. Even the "official" wisdom of the

INCOSE papers needs to be questioned, from time to time. Translated, that might mean accepting the INCOSE handbook suggestions 98% of the time, but questioning the other 2%. This might well be looking carefully at the INCOSE study results with respect to basic definitions (such as what is a system, and what is systems engineering).

Another domain for possible questioning might be new initiatives from the Department of Defense (DoD) community. An example might be any new positions (or directives) in the acquisition system. There are a large number of companies that would be affected by such new positions. You might wish to be ahead of these enterprises and be prepared to find new approaches and solutions. Another source for ideas in this connection is the chart, in chapter two, of items that are inside and outside the box.

Back of the envelope

Here we are encouraging the systems engineer to trust his or her experience and intuition in terms of finding the right and "short" path to the first iteration solution to a systems engineering problem. In this connection, one can envision team meetings in which key designers have an opportunity to express their ideas, especially regarding system architectures. This is the key element of systems engineering, and it's clearly important to get it right. We then envision a meeting in which the inputs from the key designers are synthesized by the team leader. This becomes the first-order approach, especially to the system architecture, using the experience and intuition of the most experienced designers. This approach can be modified later, as the team moves on in terms of the next formal steps of systems engineering.

Visualize solutions

In chapter two, we briefly examined the notion that visuals play an important role in developing solutions to systems engineering problems. Here we are referring to graphs, charts, pictures, and the like. It was also pointed out that using these visual approaches are definitively a cognitive activity. Indeed, they convey more information than trying to use text descriptions. So the suggestion here is to look at increasing the number of times that you use visuals, and rely on them more to make your points when addressing systems engineering tasks.

We also note here that one of the comments of Albert Einstein, regarding his approach to thinking and problem-solving, was that he leaned heavily upon his ability to "see" and "visualize" aspects of the matter. Many other critical thinkers have made the same point—they always try to "picture" the meaning of the formulas they were considering. This can be a valuable as well as more successful approach that's high on the list of both thinking and systems engineering.

Remove constraints

The problems faced by the systems engineer are quite often beset by lots of constraints. That's the nature of the occupation, so here are some of those constraints that need to be challenged, and ultimately "removed":

- Not enough funds
- Not enough time
- Poor match in terms of personnel skills
- Have right people, but not enough of them
- Lack of the right facilities

Not all constraints can be overcome, but many of them can, especially if one is able to tolerate the discomfort of being "outside the box."

Consider the EAM procedure for architecting

The Department of Defense Architecting Framework (DoDAF) approach to architecting a system is likely to be continued as a result of the commitment made over the years by the Department of Defense. Despite that apparent fact, the Eisner's Architecting Method (EAM) approach will provide a broadening of your considerations, hopefully leading to a greater likelihood of success. Recall that this involves designing several alternative systems and selecting the best among them on the basis of a cost-effectiveness comparison. Thus, the basic steps are as follows:

1. Functional decomposition
2. Synthesis
3. Analysis
4. Cost-effectiveness evaluation

Thus, the EAM approach does not replace the DoDAF procedure but rather serves to complement that procedure.

Avoid GroupThink and improve group problem-solving

Group design and problem-solving are an important part of what systems engineers do to build a system. Too often, these approaches are flawed by nonproductive and even dysfunctional processes known as groupthink. There can be many factors that lead to groupthink. Today's challenge for the systems engineering community is to find new and better ways to combat groupthink. This is often approached as a team-building as well as a leadership issue. This author would suggest that it is both, and therefore not an easy problem to overcome. A possible place to start might be to address these questions:

- Recognition—do we have a groupthink problem?
- What's the nature of that problem?
- Is the team leader capable of dealing with the problem?
- Do we need to change leadership?
- Do we need to add new people to our team?
- Do we need to remove some people from our team?

Embrace Rechtin's heuristics, including simplified design

In chapter three, we have cited, emphasized, and reiterated the contribution that Rechtin's heuristics have made to systems engineering. We envision a process whereby the key systems design team has a meeting, or several meetings, considering each and every one of these heuristics, asking the question— are we giving due consideration to this idea and its implications? In this way, the heuristics can serve as an overall guide to the system design, development, and operation.

Beyond that, special consideration should be given to simplifying all aspects of the design, using Rechtin's KISS (Keep It Simple, Stu..d) principle. This is one of the great challenges of today's world—reducing complexity and yet still achieving the results that we want.

Integrate "stovepipes" only when it is cost-effective to do so

We have seen guidance, in various parts of the federal government, that we should "integrate all stovepipes." This guidance is certainly well meaning, but can lead to significant trouble. This author has seen several programs that have failed in this attempt—were not able to integrate all stovepipes and also stay within reasonable and initial cost and time constraints. The basic problem is, in many cases, that the systems were not designed to be integrated. A classic issue is that of the software design and configurations. So—the bottom line is that integration is desirable, but we should do so only when it is provably cost-effective to do so.

Accept earlier suggestions in chapter three

In chapter three, we find ten "closing suggestions for the systems engineer." Some of these are the same as several of the notions in this chapter. However, some are different. So it makes sense to double back to those that are different and bring them into our thinking about what to emphasize in terms of what the systems engineer does. Here is a list of new and different suggestions:

  a. **Pay attention to the details of functional decomposition.** We have certainly concluded that functional decomposition is a critical part of the design process. Be careful about how this is done, and strive to be exact with an overriding rationale. Also, do not decompose to too many levels. In most situations, three levels are appropriate. For systems of systems, more levels are typically preferred.
  b. **Consider designing for both low cost and best value solutions.** The notion of three "solution" domains was introduced in this chapter. This is an approach that can provide consistently positive and successful results. Embracing this approach can lead to becoming leaders in the highly competitive acquisition world. It is also consistent with designing several alternatives as per the current DoD analysis of alternatives (AoA) approach discussed later.
  c. **Insist on developing alternatives as part of your design processes.** This commitment is supported across the board by none other than the Department of Defense. They took the lead in defining what that means as well as how to approach it [3]. Here are some of their words:
     • The **AoA** is a documented evaluation of the performance, operational effectiveness, operational suitability, and estimated costs of alternative systems to meet a capability need that has been identified through the Joint Capabilities Integration and Development Systems (JCIDS) process.
     • The **AoA** also considers the sensitivity of each alternative to possible changes to key assumptions and/or variables.
  d. **Use systems thinking in your measurements programs.** Measurements are a critical part of systems engineering. One is constantly measuring system status and one of the key issues is how to measure "at the top level" vs. how to measure everything because we might be capable of doing so. Some of this appears as an important element of the "verification and validation" processes. It also comes into play when considering one's technical performance measurement program.
  e. **Consistently use "cost-effectiveness" as your approach to your customer.** There is a strong and persistent need to stay in touch with

your customer. But it has to be done in a positive manner. One way to achieve this is to demonstrate that your system represents a cost-effective solution. One must go beyond the claim; it's necessary to show that the system is more cost-effective when compared with other alternatives (see AoA earlier).

f. **Design to facilitate integration.** Integration is one of the key challenges of the systems engineering team, and an integrated solution is, of course, a desirable one. If the team starts with a clean sheet of paper with a new system, then it's possible to start the design with the intent of facilitating integration with the current system as well as for downstream upgrades.

This "top dozen" chapter attempts to tie the main "thinking" concepts of this treatise with some of the tasks of the systems engineer. Hopefully, this will help the systems engineer and his or her community in finding new and improved ways to approach and execute in the complicated world of building systems.

References

1. Senge, P., *"The Fifth Discipline: The Art & Practice of the Learning Organization"*, Doubleday, 1990.
2. De Bono, E., *"Lateral Thinking"*, Harper & Row, 1970.
3. See Defense Acquisition University (DAU) website.

Exercises

8.1 What are the five other ways you can think of to avoid groupthink? What can you find in the literature in terms of ways to avoid groupthink? Cite the references.

8.2 Write two pages exploring more about AoA.

8.3 Elaborate upon, in two pages, how to use systems thinking in your measurements programs.

8.4 Do you develop alternatives for all your significant design processes? If not, why not? Are there ways to improve in this regard? What are they?

8.5 Write two pages on specific ways to "design to facilitate integration."

chapter nine

Final thoughts

Thinking is carried out mostly in two domains—by the individual, and within a group or team context. They are usually quite different. In this book, we focus on the former, exploring processes and approaches. Some time is devoted to groups and what might be happening there. Both are important in terms of understanding and making progress in our society.

The individual thinker

It is relatively easy to observe the wonderful contributions that have been made by the great thinkers of today and yesterday. They are much appreciated, and we marvel at the results and what they have meant for society at large.

We see these contributions, but often have no idea as to what the thinking process was that led to these contributions. In general, there are two sources for understanding the thinking process. The first source is a publication by the contributor himself or herself. They are telling us, often, how they thought about the problem. The second source is an investigator, not the contributor, that has researched the matter and has come to some conclusions regarding the thinking process that was used. Both are valuable. And both are used in this book to try to better understand "how to think" constructively about one problem or another.

One conclusion that this author has come to, by addressing the earlier two sources, can be summarized as follows:

> Thinking appears to be all about asking and correctly answering questions; the right questions

How do we know the right questions? Answer. In general, we don't. But the contributor does.

Thus, the outstanding thinkers know what questions to ask and they press for answers. They are determined to get the correct answers, and they persevere until that occurs. That may take years of investigation, but they continue nevertheless. Mostly, the rest of us give up. The agenda for the outstanding thinkers is filled with high curiosity and perseverance.

As the thinkers get answers, they ask more questions. Is this my final answer? Do I need more data? Are more observations necessary? Is there any new literature or evidence out there?

More questions are asked and answered. Am I on the right track? Do I need to change my approach? Do I need to move to some form of "lateral thinking"?

Am I where I want to be? Am I missing anything? Am I finished?

Am I ready to construct a hypothesis? Let's try this construct? Is it complete? Is it a coherent theory? Will the answer get me to where I need to be?

So these are the key words and the key attributes that appear to be most important to the thinking process:

- Asking the right questions
- Answering these questions definitively
- Persevering
- Looking for closure; a conclusion
- The attributes of endless curiosity, intelligence, and imagination

Remember, it was Einstein that said that imagination was much more important for him than was pure knowledge in his scientific work.

Forms of thinking revisited

By way of a summary, here are some of the types of thinking discussed in this book:

1. Inductive thinking
2. Deductive thinking
3. Reductionist thinking
4. Out-of-the-box thinking
5. Systems thinking
6. Design thinking
7. Disruptive thinking
8. Lateral thinking
9. Critical thinking
10. Fast and slow thinking
11. Breakthrough thinking
12. Hybrid thinking

These are some of the options the investigator needs to consider in terms of individual problem-solving.

Here are some other ideas about how to approach this matter, not discussed in detail in this book:

1. Thinking by analogy
2. Thinking by precedent or fable

3. Thinking by breaking all the rules (Einstein once again)
4. Thinking by sharing approaches and knowledge with others (group situation)
5. Thinking by association
6. Thinking by causality
7. Thinking by metaphor

Yet another perspective on thinking and problem-solving has to do with **observing** the world with a special eye. These observations ultimately lead to conjectures about how the world and the people in it operate. This is converted into a **hypothesis** (one or more) that is then tested. This more formal approach is supported, or not, by **data.** Today's world tends to provide extensive amounts of data from which to test hypotheses. Using statistical methods in this connection is a requirement as well as the correct pathway to success.

From the geniuses themselves

Part of the answers that we seek in terms of approaches to thinking can be heard by listening to the voices of the geniuses themselves. Here are some of those suggestions:

1. Aristotle—Be a free thinker; be critical and evaluate.
2. Leonardo da Vinci—Simplicity; experience is a truer guide than words of others.
3. Isaac Newton—Truth is found in simplicity; bold guesses lead to discovery.
4. Albert Einstein—Visualization vs. words; combinatorial play; minimum number of primary concepts; imagination.
5. Richard Feynman—Make progress by proving yourself wrong; explore in irreverent and original ways.
6. Edward de Bono—Lateral thinking and six thinking hats.
7. Bertrand Russell—Mathematics has both truth and beauty; No end to trouble if we abandon our own reason.
8. Daniel Kahneman—Fast and slow thinking with System 1 (intuitive) and System 2 (deliberate).

Group thinking and behavior

The good behavior of a group and the individual perspective within a group generally differs significantly from pure individual activity and thinking. We take as an appropriate conclusion that creating a highly productive and innovative group is quite difficult, and that when you find one, try to be part of it and learn from it. A good example, discussed here, is that of the Synectics Group that addressed a variety of problem-solving tasks. We recall a few features. The group composition, for example, was a set of disciplined individuals from the fields of physics, mechanics, biology, geology, chemistry, and marketing. Emphasis was placed upon the group leader and his or her specific skills. Finally, we note the orientation toward emotional issues and approaches vs. purely rational considerations. This is not expected but it appeared to work. This author prefers an approach that integrates the emotional and the rational.

We have also cited the very large group endeavors that have been successful as a consequence of excellent management understanding and behavior. Examples have been Bell Labs, the Manhattan Project, and many of the National Aeronautics and Space Administration (NASA) programs. We need more of that and the kind of thinking that is dominant in such situations.

On the negative side, we also note the existence of groupthink, a group process that leads to poor thinking and often inappropriate results. There are several forms of groupthink, and all teams and groups need to try to avoid this type of behavior. This is considered by this author to be one of the most important issues to be addressed and solved in today's competitive and sometimes dangerous world.

Exercises

9.1 Add a list of eight "voices of the geniuses," as represented (but different from those) in this chapter.

9.2 Write a three-page exploration of groupthink.

9.3 Write a three-page overview discussion of group decision aids (software "systems") that help groups come to rational and appropriate decisions.

9.4 Explain how some of the earlier decision aids work to achieve their goals.

9.5 Find and discuss software that deals with the matter of "thinking" aids. How does this software help an individual or a group improve "thinking"? Explain the process.

Appendix A
Selected thinkers: past and modern day

A league of their own

Da Vinci—Perhaps the most productive genius ever; he excelled in art, science, and engineering, as well as just about everything in which he participated (e.g., aerodynamics, hydrodynamics, weaponry). See the seven intelligence areas suggested by Gelb in chapter six.

Einstein— Brilliant physicist of 20th century. Formulated theories of relativity, the most significant breakthrough in science in modern times.

Astronomy

Kepler—A German mathematician who developed the Laws of Planetary Motion and improved upon the refractory telescope.

Copernicus—Documented his "theory" that in fact the earth orbited around the sun, a breakthrough at that time (around 1500).

Galileo—An Italian scientist, sometimes called "The Father of Modern Physics," constructed and utilized powerful telescope; supported heliocentrism.

Science

Hawking—Leading contemporary physicist who has contributed significantly to cosmology and the theory of black holes.

Pascal—A French mathematician who developed and wrote about the philosophy of mathematics; built a mechanical calculator; developed the binomial coefficients (Pascal's Triangle).

Faraday—Developed the first electric motor, as well as generators, transformers. and overall theory of electromagnetic induction.

Darwin—His "Origin of Species by Means of Natural Selection" was a milestone in his theory of evolution, changing the way we think about this complex field.

Newton—Formulated gravitational model, along with significant contributions with the theory of calculus and other areas such as optics and classical mechanics.

Maxwell—Defined formulae for electromagnetic waves and their propagation, thus setting the stage for the development of all manner of communications systems.

Descartes—Was "Father of Analytic Geometry"; Wrote about Meditations on a First Philosophy; Well known for Cogito Ergo Sum (I Think, Therefore I Am) declaration.

Edison—Not a "scientist," but most prolific inventor dealing with technologies such as incandescent light, the phonograph, electric heat, light and power systems.

Marie Curie—Polish/French physicist who made significant advances in radioactivity; winner (twice) of Nobel prize; worked with husband Pierre as both of them are credited with breaking new ground in physics and chemistry.

Social science

Freud—Considered the "Father" of psychoanalytic theory and practice. Mentor to a large group of followers around the world.

Jung—A Swiss psychiatrist that founded analytical psychology; a leader with archetypes, self-actualization, and other unique concepts. Has substantial following today, rivaling Freud.

Mead—Significant contributor to the field of anthropology; known for "Coming of Age in Samoa" as well as substantive studies of race, sex, and intelligence.

Maslow—Formulated "Hierarchy of Needs" concept; Developed principles of self-actualization, transpersonal psychology, and human motivation.

Statesmen

Plato—One of our earliest philosophers who taught everyone how to think. His mentor was Socrates, which speaks well for both of them. Wrote "Oration on the Dignity of Man" and other classics.

Aristotle—An ancient Greek philosopher who led advances in logic, metaphysics, ethics, economics, and politics.

Marx—German political philosopher, led approaches in historical materialism, class struggles and conflicts, revolutionary actions, and Communist theories, thinking, and dogma.

Jefferson—Considered to be the primary author of the Declaration of Independence, a brilliant thinker, and the founder of the University of Virginia.

Franklin—Diplomat and scientist, as well as one of founding members of the country. Inventor of a variety of devices for home, office, and factory.

Lincoln—Credited with bringing the country through the Civil War, and for the significant Emancipation Proclamation. Considered one of our greatest presidents and orators.

Gandhi—Defined and led the "passive resistance" ideology and movement from his native country, India. Set example for Martin Luther King approach in the United States.

Locke—English philosopher, known by many as the "Father of Liberalism." Formulated various theories and writings in such areas as the mind, identity, ideas, tolerance, and empiricism.

Hobbes—English philosopher in 1600s who wrote groundbreaking "Leviathan," dealing with social contract theory; led with new ideas regarding natural equality, jurisprudence, and even certain aspects of physics.

Medicine

Mendeleev—Formulated the "Periodic Table," which was the foundation for thinking about and using chemistry for all researchers and students.

Salk—Well known for development of the vaccine for polio that bears his name.

Watson & Crick—Formulated a model of DNA structure. This was a major breakthrough for which they won the Nobel prize.

Business world

Drucker—Most cogent and prolific observer of the business scene from the academic world. Originator of widespread Management by Objectives practices.

Collins—Wrote two penetrating and influential books—*Built to Last* and *Good to Great*, describing important features of various businesses and how they have operated.

Peters and Waterman—Also observers of the corporate scene but with great insight wrote blockbuster book *In Search of Excellence*. Confirmed nine attributes of excellent companies.

Tom Watson, Jr.—As CEO of IBM, brought company into mainframe computers, and great success with 360 series. Established strong corporate beliefs, along with "think" emphasis.

Geneen—Began stint as CEO of International Telephone and Telegraph (ITT) in 1959, leaving in 1977. During that period, increased revenues from 766 million to 16.7 billion. Early purveyor of monthly measurements process.

Welch—Took over as CEO of GE in 1981. During his tenure, company increased market cap by more than $450 billion. "Fix, sell, or close" approach highly successful in terms of company revenue and profit growth.

Augustine—President of Lockheed-Martin and key advisor to government (e.g., Department of Defense, National Aeronautics and Space Administration (NASA)) on matters of national technology, security, and management. Professor and trustee at Princeton University.

Hewlett and Packard—Built technology-based company from local garage to leading "systems" enterprise. Advised the DoD on systems and security matters at highest levels.

Ramo—As the "R" in TRW, he led the charge in systems engineering, the country's ballistic missile development, along with associated military technology; advised the DoD and other high-level agencies with respect to nation's overall position and strategy.

Gates—Most significant founder of Microsoft. Brought company into Windows operating system, capturing significant portion of the market. Solid strategic thinker and manager.

Jobs—CEO of Apple and earlier versions of company. Incredible sense of product design. Brought new products to market, including iPhone, iPad, and iPod. Led Apple to achieve "cap value" of $377 billion in 2011, one of the highest at that time; grew to highest (over a trillion).

Bezos—CEO of Amazon, breaking new ground in revenue levels and approach to product expansion as well as services to customers.

Appendix B
Suggested topics for a first course in "STEM"

There has been much discussion of the need for and relevance of STEM—science, technology, engineering, mathematics—in terms of educational systems across the United States, and indeed, the world. In this Appendix, we suggest a "first course in STEM" at the precollege or first year of college level— topics for a three-credit course that fits into existing programs or serves as a "starter" for STEM. Consider it a *gedanken* experimental list to get some "thinking" started, if such is on one's agenda. There are some 42 "contact" hours that might be associated with 3 h of lecturing each week for 14 weeks, a typical (but flexible) course that may leave some time for joint projects and/or examinations.

Topics within 3-h session for first STEM (three credit) courses	Cumulative hours
Week 1—Thinking, Advanced Algebra, Integral Calculus	3
Week 2—Differential Calculus, Astronomy, Cosmology	6
Week 3—Atomic Particles, The Periodic Table, Gravitation/Forces	9
Week 4—Important Seminal Technologies, Modern-Day Technologies, The Computer	12
Week 5—Key Computer Science Ideas, Key Computer Applications, The Spreadsheet	15
Week 6—The Database Management System, Innovation, Companies and Laboratories (Bell, Xerox Parc.)	18
Week 7—Classical Physics, Electricity & Magnetism, Thermodynamics	21
Week 8—Great Science Ideas, Inorganic Chemistry, Organic Chemistry	24

(Continued)

INDEX

Printed in the United States
by Baker & Taylor Publisher Services